Basic Factory Accounting and Administration

Basic Factory Accounting and Administration

M W Monaghan

Gower

658.5
M73 b

Published by
Gower Publishing Company Limited
Aldershot, Hants, England

m *

British Library Cataloguing in Publication Data

Monaghan, M.W.
  Basic factory accounting and administration.
  1. Factories — Accounting
  2. Factory management
  I. Title
  658.1'5   HF5686.M3

  ISBN 0–566–02281–8

Typeset by Inforum Limited, Portsmouth
Printed in Great Britain by
Redwood Burn Limited, Trowbridge, Wiltshire

Contents

Preface

This book is intended primarily for busy managers and accountants, and those aspiring to these positions, mainly in the countless smaller and medium-sized industrial units. Its aim is to offer a selection of not too sophisticated ideas, from which those most appropriate to the needs of the concern in question can be taken and adapted, as required, to further its development.

Many managers and accountants in the larger undertakings, of course, will have no need of this book, as their organisations will have their own sophisticated structures in being, tested and proven. Nevertheless the subject of finance for the non-financial manager is perenially topical, and some interest may well be aroused among such individuals even in the bigger companies.

In the smaller and developing concerns, however, the case is more clear cut, and it is felt that there is a compelling need for a simple accounting guide, with an outline of factory organisation etc. to provide background. The book even contains a very basic opening chapter covering overall policy. No apology is offered for the fact that it is basic, as overall policy often has to be rethought many times during the life of a business, but at no time is this likely to be more important than in the early stages of development, such as when a small business with growing pains is on the threshold of expansion.

In general, the structure and coverage of the book reflect the routine activities of the factory, and a separate chapter has been allocated to each of the main functions. A number of topics – including quality control, factory maintenance, and personnel and training – did not seem to me to justify separate chapters, and these are dealt with together in a chapter called 'Ancillary Services'. Because the accountant will normally need to know something of the requirements of a sales organisation, a chapter on selling has also been included.

In the very small firm some of the functions described in this book may not even be carried out. In other cases, two or more functions may be undertaken by one individual, and there will almost certainly be greater flexibility generally in the smaller firm concerning this question of the allocation of functions.

It would of course be a futile task even to attempt to lay down rigid guidelines for every kind of manufacturing organisation. There is such a vast variety of industrial enterprises in modern society that it is best to regard each business as unique, and to structure its organisation to meet its own special needs. Nevertheless there are certain factors common to all manufacturing units and it is my hope that within these pages the manager and accountant will find something that is useful to them in their jobs, and in their attempts to improve business efficiency. The book also offers, I suggest, something to trainees finding their way forward in this kind of environment, in helping to bridge the communications gap that too often exists between different job disciplines.

M.W. Monaghan

1

Overall Policy

OBJECTIVES

The first, albeit elementary, step in setting up a new, or acquiring an existing, business is to establish clear objectives. What is to be made to produce profit? And on what basis? Are goods to be manufactured for stock and advertised and marketed by the company, or is the manufacturing effort, or any part of it, to be employed in producing to customers' orders? Will there be repeat orders or will orders be taken on a one-off basis? Will long-term contracts be entered into and how much work, if any, will be carried out off the site, whether at the customer's premises or elsewhere? To what extent will sub-contractors need to be engaged on operations which the company has insufficient capacity to undertake itself? Apart from work unsuited to the company's operations, and yet forming part of the finished product, jobs could be sub-contracted; for instance, in cases where the company, as a matter of policy, decides not to employ labour of the class required or to acquire plant of the sort needed for the work to be carried out.

These primary decisions will involve making an assessment of the market and the likely demand for the company's product, taking into account the strength of competition and the strategy to be adopted in countering it. Consideration will also have to be given to the likely life of the product or products, and the effects of obsolescence and possible changes in demand caused by fashions etc.

From these initial considerations, even though they will undoubtedly be modified in the course of ordinary development over a period of time, everything else follows: the nature of the organisation to be set up; the accounting methods to be adopted and the possible scope to be allowed for expansion or elaboration of both, where applicable.

RESPONSIBILITIES

A company has many responsibilities to bear in mind. Paramount among these are its responsibilities to customers, employees and shareholders. The company's policies must therefore be so planned as to keep what it considers to be a proper balance between them.

The company's customers are entitled to the best quality and service possible commensurate with the price of the article produced, and that price must be fair to the company as well as the customer. It is a mistake to cut the price to the point where the service to the customer suffers, and although price levels must be determined by management in the final analysis, the accountant must provide the basic data to assist in this piece of decision making. As to quality and service, these will depend to some extent on the product. It would obviously be as ludicrous to use high quality materials where the need did not exist, as to use inferior material for a high quality product; just as it would be foolish to despatch a non-urgent consignment by air, when assuming suitable weight and volume, it could be conveyed by a less expensive method later.

Employee policies should be devised to promote optimum productivity, having regard to preserving equity among the various grades of labour in the light of local conditions. Working conditions should be as good as practicable, if only to further this objective, and special consideration should be given to the safety factor. Employees' contracts should be as unambiguous as possible to minimise the risk of disputes and bad feeling in the event of possible misinterpretation. In its own interests the company's responsibilities to its employees should not end at the place of work, but should take into account such social and welfare benefits as can be afforded consistent with trade practice. A duty is owed to employees to see that they are aware of what job changes, long-term prospects and promotion possibilities they can reasonably expect in normal circumstances, and the time and means must be found to maintain communication to reduce the risk of grievances building up and the company losing the initiative.

It goes without saying that the company has a commitment to the providers of its capital, who placed their funds at risk, enabling the acquisition of assets to be made for the company to operate, and from which operations the company would hope to make its profits. In the event of a loss, of course, the company cannot pay anything to its shareholders, except out of past profits, whilst sustained losses would sooner or later force it to cease operations. For a company making profits, the ideal state of affairs would exist when its customers were being charged fair prices, its employees being paid fair wages and salaries and its shareholders receiving a fair return on their investment. The point should not be overlooked that, except in times of extreme adversity when the very livelihood of the wage and salary earners may be threatened, the shareholder suffers by comparison with employees in so far as wages and salaries are paid whilst the company operates, whether profitably or not. Therefore it is fair and reasonable that, whenever the circumstances afford, higher dividends should be paid following a period when a low, or even no, dividend was paid. There are bound to be arguments at times over how the cake is divided, but much unnecessary trouble can be avoided by good public relations policies demonstrating the company's endeavours to preserve fairness.

LOCATION

The choice of location will take into account such factors as the cost and availability of premises; local amenities; the degree of shortage or superfluity and the market rates of the classes of managerial, clerical and shop-floor skills in the area, and the means that might be available to attract more if needed; transport facilities, road, rail, sea and air, as appropriate; distance from potential customers and major suppliers; national and local government plans for the region; and scope for expansion in the area.

A commercial approach on the part of the factory manager or accountant, involving consideration of all these factors, will optimise the input to the decision-making process and reduce the risk of serious miscalculations being made.

PREMISES

The choice of premises is, of course, closely linked to the choice of location, and the circumstances of the case will determine which is to take precedence. It may first be decided to set up operations in a particular area and consequent to this decision to search for, or build, suitable premises within that area. In other, perhaps more specialist, cases the choice of premises might be the primary consideration and a specification given to estate agents, the choice of location being a matter of secondary importance. Apart from the matter of occupancy costs, a number of other factors, some technical, will have to be given consideration varying in importance according to circumstances. Are there any special requirements regarding heating or air-conditioning? What parking facilities are required? Will the premises need strengthening in any part due to the operation of heavy machinery? Any special head-room requirements? Do sanitary arrangements etc. comply with the appropriate regulations? Are the premises in the proximity of a residential area and, if so, are there any precautions to be taken with respect to noise and pollution? If the premises are to be leased, the lease will be studied and due note taken of any clauses or schedules dealing with dilapidations and periodical decorations.

PLANT

When the decision has been made as to what is to be produced by the company on its own premises, the plant including ancillary equipment, with which to achieve the required production must be studied. In practice, of course, the decision as to what is to be produced will, in itself, often be influenced by the plant factor. It may be preferable to purchase some items of plant second hand and this may depend on what is to be produced, the anticipated length of the production run and funds available. In this connection it may be decided that some of the more expensive items should be leased. If special purpose, i.e. custom-built, equipment is needed to produce certain parts or carry out certain processes, the anticipated production life of the project in question must be looked at to see over what period the plant is to be written off, and consequently to consider what special depreciation provisions should be made. Technical management will decide on the plant layout, of course, in line with the production flow where feasible; and the question of what ancillary equipment is needed, e.g. between processes, such as fork-lift trucks, conveyor systems, special lighting etc., will be looked at in terms of cost, safety and productivity.

Servicing time and costs must be allowed for and in establishing the plant requirement, a contingency should be built in for normal breakdowns and repairs to safeguard the factory output.

TRANSPORTATION

Certain factors require consideration before a decision is made on the mode of conveying the company's goods: distance from potential customers; nature of goods to be carried, i.e. weight, volume, fragility, packaging; volume and regularity of consignments to be made; storage, garaging, servicing and loading facilities available; distances from airports, seaports, motorways and railheads; the degree of urgency likely concerning deliveries and, of course, availability of funds. In the vast majority of cases the goods will at least start their journey from the factory by road transport, but it may be necessary to transfer later if the journey is to be continued by air, rail or sea. With modern methods of containerisation, however, it is not unusual nowadays for goods to reach a customer on the continent of Europe in the same lorry as that in which they left the factory.

If deliveries are likely to be spasmodic, it might well be worth hiring a carrier as required, rather than incur the expense of maintaining company vehicles and transport staff who are not fully employed during the slack times, and possibly not able to cope with the load during peak periods. Obviously, the aim should be to achieve maximum utilisation to reduce transport costs as much as possible. For instance, with proper planning to avoid unnecessary waiting time etc., it might be feasible to arrange for goods to be collected from a supplier after making a delivery to the customer, thus saving an empty-handed return journey.

WORK FORCE

When establishing a manufacturing unit in a given locality, a number of points regarding the work force have to be taken into account. These will include the question of the degree of scarcity or otherwise of the type of skills required; whether there is a need for special training and, if so, whether suitable trainees can be attracted. In this connection the training facilities required by the company might be determined by whether or not there is a suitable training college or government training centre in the area. Remember also that training costs can be reduced by the grant claimed, the amount depending on the quality and extent of training. Rates of pay in the district for comparable jobs must also be looked at as well as other special factors, such as the background to any history of industrial unrest in the area; prospective competition in the labour market from other employers if the area is in course of development etc.; local economic conditions and, if there is a lack of certain amenities, whether special provision need be made to cater for the needs of the workers. Obviously, the cost of any such provisions must receive consideration. In deciding whether or not the company can afford expenditure of this nature — such as the hire of coaches to carry workers to and from the factory, or the provision of a sports ground — the possibility of sharing the costs with other employers should be studied.

For example, the factory might be mid-way between towns 30 miles apart, and it might be practicable to collect, say, a coach-load of workers from a town or district 15 miles to the west, and another 15 miles to the east, and whilst one coach might be fully occupied by the company's employees, the other might be only half full. Another employer in the district, working the same hours, might have half a coach-load and the cost of running that coach could be shared, being reviewed periodically as circumstances changed, of course. Equally, the extent of any sports and social club subsidy the company might make could be influenced by the prospect of sharing facilities with another employer.

FINANCING

It is normally preferable for the company to be self-financing as far as possible. The smaller company, naturally, will not have a Stock Exchange quotation, and the share capital structure will not usually be, or indeed need to be, very elaborate. The amount of issued capital, and its division into shares preferential as to dividend and ordinary shares will probably depend on the need to attract capital and the sources available. The reason why it is not usually desirable for outside interests, i.e. other than shareholders, to provide finance, on too large a scale at any rate, is that the larger the stake they have in the company, the more costly the finance charges, and consequently the harder it becomes for the company to pay dividends to its shareholders.

Nevertheless there are times when most companies, large or small, need extra finance for their operations, without wishing to issue more shares. The smaller company, when commencing operations, may wish to restrict the number of shareholders in order to give them a greater share of the hoped for expansion when the outside interests, e.g. debenture holders, are paid off. The question will usually revolve around renting or buying premises or plant. If plant is not purchased outright by the company, it could perhaps be acquired on hire purchase, or possibly it could be decided to forsake ownership and merely to lease the equipment. A bank loan or overdraft might be negotiated, and this could be used to finance the stocks and work in progress that the company would need to carry. The amount of finance necessary will depend chiefly on the size of the operation and the length of the production cycle, and the choice of method of financing will centre usually on the sources available and cost.

2

Factory organisation

PERSONALITIES

Man's individuality is, arguably, his most notable characteristic. A great deal of study has taken place as to why one man, or woman, is more articulate, or diplomatic or intelligent than another; or why one person inspires more loyalty or respect than another; or even perhaps why the same person displays more intelligence regarding some aspects of his job than others. Behavioural scientists have reached, and will continue to reach, many and varied conclusions, but as far as our subject is concerned we must simply accept that owing to the diversity of the human race no two people will ever carry out a managerial or supervisory job in exactly the same way in every detail. We must, therefore, reflect on how this fact is to be reconciled with the idea of a continuing organisation with a permanent structure, despite probable personnel changes in the command chain over the years.

The first thing to be realised is that the organisation itself is a living thing, and that, therefore, its structure is bound to change anyway, to a greater or lesser extent, as it develops. It is inevitable that changes of individuals will often bring about some modifications to the organisational structure, perhaps not in job titles but in the amount of work delegated or responsibilities covered. Nevertheless, a balance has to be struck in order to achieve a structure that will be disturbed as little as possible when personnel changes occur. Although it is perfectly natural that jobs and responsibilities may change according to whether one occupant has a greater capacity for work than another, or perhaps a more efficient method of working, the organisation itself should be so designed that it does not have to be changed too radically. Drastic changes invariably cause more work and inconvenience, are upsetting and can do lasting damage.

COMMUNICATIONS

It is important to ensure that sufficient time and effort are expended so that everyone concerned is aware of, at the very least, that part of the command structure with which he is directly involved, and consequently of the limitations, if any, imposed on the responsibilities of the personnel therein. This is not to suppose that rigid lines of demarcation can be drawn for every case, but the aim should be to reduce to a minimum those instances where a person is faced with a choice and forced to decide which of two or more persons or departments should be approached. To give some permanence to the matter and to avoid misunderstandings it is desirable to have the necessary information transmitted on paper. This provides continuity and saves time-wasting repetition of verbal notification when changes of personnel occur, and of course enables easy reference to be made in cases of doubt. Naturally, the foregoing comments apply as much to outsiders dealing with the company as to personnel within the company itself.

To take a simple example it is a good idea, for instance, to list the functions of executive directors on the company's notepaper, thereby enabling, say, the purchasing manager of a prospective customer to know the name of the sales director, should it be necessary or appropriate to establish individual contact. Within the company, the best way of communicating functions is undoubtedly by the use of organisation charts, which are touched on later in this chapter. It will be apparent that lack of information regarding the chain of responsibilities, whether entirely within the company or dealing with outsiders, can only lead to confusion and inefficiency.

FUNCTIONS

It will usually be found that the clearer the conception of functions, the smoother the organisation works. While it is advisable that as many aspects as possible of a manager's, supervisor's or clerk's job be covered by one or more other persons in case of necessity in a comparatively small undertaking, it is nonetheless essential to know precisely whose primary responsibility it is to see that each function is satisfactorily performed. It might be that the planning engineer also carries out the estimating and work study functions, and in his absence temporary cover might be arranged by the draughtsmen preparing estimates, the production controller undertaking planning and the works manager carrying out work studies. However, in such a case the planning engineer would remain the person with chief responsibility for these functions on his return.

Of course, as the organisation develops, it may be necessary to rearrange functions as circumstances demand, and this should be done in a clear-cut way so that all concerned are aware of the new arrangement. Our planning engineer, for example, could have some spare time on his hands and it might be decided to give him responsibility for the progress function too; on the other hand, if the workload increased he might well lose the estimating and work study portions of his job, and indeed it could even be necessary to engage another planner to assist him. A certain amount of elasticity might have to be allowed. In some cases, perhaps, one person carries out more than one function, but in others one function is carried out by more than one person. The vital point for management to consider is that every function should be catered for, and where one person does carry out more than one job, it is good sense to see that some form of internal check is built in, particularly if the jobs are related, to help prevent unrectified errors. For example, this could perhaps happen if a buyer was also responsible for stock control. It can be very useful to prepare written job specifications, detailing the individual's duties and responsibilities, together with the relationships to other functions but, in view of the foregoing comments, room should be left for amendment.

ORGANISATION CHARTS

The organisation chart should not be cluttered up with too much detail, and the larger and more complex the organisation, the more important is this principle. If necessary, subsidiary charts can be used to fill in details not properly covered by the master chart. Naturally, both names and functions should be shown on the chart, but the outlining and grouping of functions is more important than names or job titles. Broadly, functions could be divided into the following areas:

(a) research and development;
(b) production;
(c) quality control;
(d) selling and distribution, and
(e) administration.

These in turn, of course, are subject to further subdivision. For instance administration could include:

(a) accounts;
(b) secretarial;
(c) organisation and methods, and
(d) personnel.

The reason quality control has been suggested as a separate function from production is that the quality controller usually reports directly to the managing director or general manager, and not to the works manager, who is responsible for production. It will be well appreciated that there can be no standard form of organisation suitable to every kind of manufacturing concern, due to the vast variety of industrial undertakings, even in the same field and between units of the same size, and for this reason the sample charts shown are only intended to illustrate possibilities.

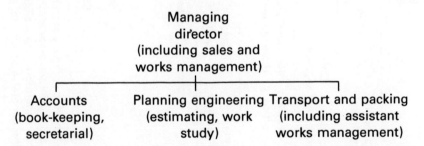

This simple chart pictures an extremely small undertaking in which the executive and administrative functions are virtually all carried out by just four people. A more elaborate chart such as might be associated with a larger enterprise is shown on p. 12.

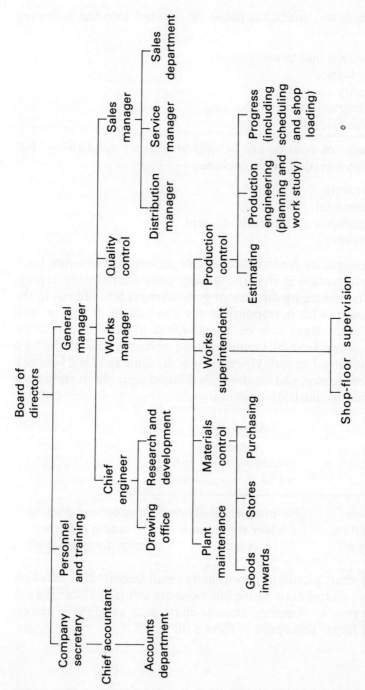

Board of
directors

Company secretary — Personnel and training — General manager

Chief accountant
Accounts department

Chief engineer
— Drawing office
— Research and development

General manager
— Works manager — Quality control
— Sales manager

Works manager
— Works superintendent
— Production control — Estimating — Production engineering (planning and work study) — Progress (including scheduling and shop loading)

Materials control
— Plant maintenance — Goods inwards — Stores — Purchasing

Works superintendent
— Shop-floor supervision

Sales manager — Distribution manager — Service manager — Sales department

Note: The reason that the company secretary is not shown as reporting to the general manager is that he is secretary to the board. As the secretary is usually responsible for the company's accounts, the chief accountant and his department are therefore responsible to the secretary. The chief accountant would nevertheless have a responsibility to the general manager to provide him with up-to-date management reports and relevant information.

3

Product development

PURPOSE

The chief purpose of product development is innovation or, in other
words, the application of fresh ideas leading to product improvement
or substitution: or the introduction, within whatever limits may be
imposed, of an entirely different product range. For the company
manufacturing its own proprietary products, and to a lesser extent
even the company engaged in jobbing or sub-contract work, it is
tremendously important to have an active 'think tank' in the field of
product development, so that the company can at least keep abreast
of its commercial competitors.

To be economically advantageous, research into new products or
processes should naturally be carried out on a scientific basis, as
opposed to in a random or 'hit and miss' way. The extent to which
funds can be committed to product development depends, as indeed
with the allocation of funds for any other purpose, upon management's assessment of priorities in the light of available resources. One
way is for a specific sum, perhaps, but not necessarily, based upon a
certain percentage of the previous year's profits, to be allotted as the
research budget for the year and for the head of research to be given
certain guide lines as to the path along which his efforts should be
directed. The limits imposed will be contingent on the circumstances
of the case, but within these limits the research head must use his
discretion as to what is technically and commercially feasible, and he
will usually have occasion to consult with his colleagues within the
company to obtain from the experts in their own fields information

and advice from time to time. This may relate to technical matters or marketing possibilities, whilst the accountant may be required to provide cost data in connection with possible alternative materials, sources of supply, materials usage or wastage, different or speedier processes, or modifications to existing products to improve safety or durability. Such data will obviously assist the research head in arriving at his decision on the course to be adopted.

In any event, when research spending is involved, the accountant should provide, as part of his periodic management accounts, details of costs to date compared with budget. If certain sums have been allocated to specific projects, the research costs statement can be set out to show the amounts budgeted and cumulative costs of each, if arrangements are made for projects to be numbered. The project number can then be shown whenever expenditure is incurred so that the costs can be booked against it.

PRODUCT DESIGN

A good research and design unit avoids a 'head in the clouds' outlook, and instead keeps an ear to the ground to pick up the reaction of others to the results of its work. An effective feedback system helps to keep it informed of consumer and specialist opinion and can possibly, therefore, provide the company with an advantage over its competitors by enabling design faults to be found and corrective measures taken. Although this may appear too obvious to mention, there have been many outstanding cases in which the designer has seemed to be unaware of consumer demand or of another technical expert's requirements.

A contention that received some publicity during the 1950s was that houses were designed by men and that too often the housewives' needs were not taken into account. Consequently house designers were made aware of this fundamental omission and improvements followed: for example, in such things as kitchen layouts and cupboard space. Many garage mechanics will assert that cars are invariably designed to make the engine as inaccessible for servicing as possible. The easier a product is to service or repair the cheaper its maintenance costs, and the company is, therefore, able to use this as a selling point.

Unless a product is purely ornamental, precedence must be taken by the practical, or functional, over the aesthetic considerations. The importance of the latter, however, must not be minimised, as it will be apparent that consumer demand can be influenced by the affect of the product on the senses, particularly by its looks. Style is, therefore, a principal factor as far as the end user is concerned, but this will be balanced against practical considerations. A good-looking product that does not work properly stands little chance against a plain but practical one. This question of manufacturing design is thought to be of sufficient importance for the State to have set up the Council of Industrial Design, and industrialists are able to avail themselves of the facilities it offers for information and advice in this field.

Design may also enter into the packaging aspect where appropriate, and good designing here both protects and helps to sell the product. Other practical aspects the designer must study are:

(a) safety of both workers and consumers;
(b) stresses and strains likely to be suffered by the product during its normal useful life — experiments and tests simulating as far as possible the likely conditions to be encountered can help in this regard;
(c) standardisation. Cost benefits can be achieved in many cases by standardising machines or materials. For instance, a company making two products with holes in the centre may find that both holes can be made the same size, and this might therefore be possible using the same tool to effect the operation;
(d) environmental – this means taking into account such things as pollution, noise, wastage, etc. arising from use of the product and overcoming them or reducing the risks, exemplified by the positioning of diesel train exhaust pipes carrying the fumes above the train's roof level to be dispersed.

TOOLING

Whatever the title of the technical head concerned, be it technical director, chief engineer, or something else, his department will be concerned not only with the goods to be manufactured, but with the appropriate tooling with which to produce the goods, as it would be plainly ludicrous to design a new product without paying regard to how it was to be made. Therefore, preparation of the tooling programme for submission to management is usually his responsibility, involving an assessment of the anticipated production volume, the time scale, and an appraisal of suitable machine tools, jigs, fixtures etc. From this information together with such technical cost data as may be required, possibly based on estimates supplied by the buyer, management must decide as to how the total tooling costs are to be recovered. This may, for instance, be done by dividing the tooling cost by the number of items expected to be produced, thus arriving at a tooling cost element per component and adding this cost to the proposed selling price; or it could be decided to sell the tooling to the customer, if the parts are to be made for one buyer, whereupon on production of acceptable samples the tooling costs could be billed complete, and the tooling, although remaining in the possession of the manufacturer in order to produce the parts required, becomes the property of the customer.

From the accountant's angle the treatment to be accorded to the tooling costs is of vital importance. Obviously from the point of view of keeping his books straight, no problem arises if the customer is to be billed with the total on or before commencement of production. In other cases, however, his management may decide that, particularly where the anticipated production volume appears to be indeterminate, the cost should be written off by equal instalments over an agreed period. Perhaps more commonly though, when an assessment has been made of the likely production as indicated, management's decision might be to write off to cost of sales account that portion of the cost applicable to the quantity of parts produced and to credit the tooling amortisation account. The obvious danger in this case is that the exact amount of the total tooling cost may not be written off in this way unless, either revisions are made during the course of the production run, or a tooling amortisation adjustment is made at the end of it.

TECHNICAL RECORDS

The chief draughtsman is often responsible, inter alia, for the collation and updating of technical records. Literature emanating from outside sources, such as trade magazines and British Standards Institution publications, can be carefully catalogued and a technical library formed so that articles and data of interest in connection with a particular project can be referred to as and when the need arises. The parts making up an assembly are scheduled and a parts list built up, referring to the drawing number and quoting part numbers and quantities required. From this information the materials controller can determine stock levels and implement his purchasing programme.

A sensible product numbering system can save considerable sums in terms of avoiding wasted time in ascertaining material types, locating product ranges applicable to the parts in question, determining sources of supply etc. All sorts of permutations can be used, but a simple example would be as follows: the first two digits to signify the model or assembly number, with sufficient allowance being built in to accommodate common parts, the third to indicate the material type, the fourth indicating colour, and the fifth and sixth digits relevant technical data, e.g. dimensions and tolerances. In some numbering systems an indication of the supplier concerned is sometimes incorporated, but where this is done care must be taken to see, for instance, that parts are not ordered needlessly, as could happen if someone was unaware that the identical part was already available, having been supplied by another firm and given a different part number.

All this naturally implies the drawing up of code lists or charts for each category, reference to which would provide the key to the product or part number.

4

Selling

MARKETING PLAN

Before anything can be manufactured to produce a profit, there must be in the very first instance an estimation of its selling potentialities. This implies a conscious assessment of (a) likely markets; (b) quantities that can be sold; (c) selling prices; (d) possible outlets, e.g. agents, in addition to, or in place of, the company's own sales force and (e) methods of packaging and distribution. Unless monopoly conditions apply, most, and in some cases all, of these factors will also involve considering the competition likely to be met, and the resultant answers will obviously be coloured by whatever influence this competition is thought to be capable of exerting.

The final answers, suitably tabulated, can form the company's sales budget, subject to modification by management, if necessary, to take account of manufacturing capacity. For example, to make full use of capacity on one group of machines, it might be decided that the sales budget for that particular item produced could be increased, and the extra parts sold possibly with the help of additional advertising. On the other hand, it could be decided, due to scarcity of the kind of skilled labour required, to scale down the budgeted sales of another component, and accept a more realistic figure. However, once the budget has finally been sanctioned, important though it is, it is only a plan, and it would be wrong to attempt to follow it slavishly to the point where the planners are blinded to other possibilities. The sales manager must keep abreast of developments, especially during the later stages of the plan, so that he is aware of major changes of circumstances since the formulation of the plan and can make or suggest appropriate adaptations.

SALES PROMOTION

The form and content of the company's advertising will vary according to the nature of the case, and it is both fruitless and impracticable to view sales promotion from an archetypal standpoint. There is, however, no doubt that considerable expertise is required if an effective job of sales promotion is to be executed. A knowledge of markets, both existing and potential, is essential and of the most appropriate means by which these may be reached. Obviously, a washing-machine manufacturer would not advertise his product in the same magazine as a machine tool producer; one would use a consumer paper whilst the other would be more likely to advertise in a trade journal. Where media advertising is concerned, such things as television audience ratings and trade journal circulations may repay considerable study by the advertising man.

The end in view is to increase output and profits and to reduce unit costs, and the aim of sales promotion can be defined as laying the groundwork by which as many people of the right categories and in the right areas acquire as much knowledge as is necessary to enable them to make an eventual decision to buy the company's products. The company may employ its own specialist staff to deal with sales promotion or it may entrust this area to an advertising agency. In either case, as in any other field, control must be exercised over expenditure and an appraisal made of results achieved, and in this the accountant can play an important part by correctly analysing the figures in his accounts.

An arbitrary form of analysis would be to group expenditure under the following heads:

(a) sales promotion salaries;
(b) media (press, poster, TV etc.);
(c) demonstrations and exhibitions;
(d) sales literature, and
(e) packaging (i.e. 'brand' advertising, such as highlighting the product by the use of specially printed cartons etc.).

This could, of course, be expanded if necessary to allocate these costs over the different products, where more than one product is manufactured.

AGENTS AND REPORTS

A company might decide to appoint an agent or agents to sell its products, instead of or in addition to employing a sales force, and a decision of this sort could be centred around a particular product or geographical area. For instance, a company manufacturing both industrial and domestic electrical switchgear might prefer to establish good contacts with wholesalers to market the domestic product, but to employ its own sales staff to sell the specialist industrial equipment. In another case, rather than organise his own sales network in the North of England and Scotland, a West Country industrialist might decide to appoint distributors to handle his sales in those areas, while his own salesmen look after the remaining territories.

Having established what is to be sold, as well as how, where and when, an effective monitoring system should be organised so that results can be compared with the marketing plan and early remedial action taken when appropriate to improve performance. The agent's or salesman's reports will provide the basis for this information feedback. The format may vary considerably, and in all probability the salesman's report will be made more frequently and in greater detail than that of an agent, mainly due to the fact that the salesman is an employee and the Company will wish to keep itself informed of his whereabouts and activities. However, such reports might include the following items: (a) details of orders obtained: (b) name of customer/prospect: (c) assessment of possible sales volume; (d) details of competitors' products, if any, purchased or contemplated by prospect; (e) chief barriers, e.g. price, delivery, technical factors etc. and (f) further product information requested.

SALESMEN'S REMUNERATION

Another important factor to be considered is the scale and method of the salesman's remuneration and expenses. Management must decide if commission is to be paid and the basis of calculation, whether expense allowances are to be paid or, instead, if out-of-pocket expenditure is to be reimbursed. Is the company to buy or hire cars for its salesmen, and is private use to be paid for? The frequency of payments to salesmen – weekly or monthly; the method of payment – cash, cheque or credit transfer and whether or not expense payments can be combined with payments of salary or commission – will largely depend on the nature of the organisation. The accountant must be prepared to advise on these matters and to organise his sphere of activity accordingly.

CHARGING MARKETING

It is sometimes thought advisable to separate the manufacturing and marketing functions completely in the accounts, as a means of judging the performance of each division. Indeed, this is sometimes taken to the point of forming a separate company to handle marketing, the manufacturing company then charging marketing with the completed products to be sold, leaving the selling organisation to earn its profit from the mark-up it applies to arrive at selling prices. Where this separation occurs, whether or not involving the formation of another company, the most important factor concerns fixing a fair transfer price. Obviously there would be no point in transferring on a cost plus basis, as this would in no way serve as a measure of manufacturing efficiency. Predetermined standard prices at some point between estimated manufacturing costs and proposed final selling prices are usually chosen. Whether or not such a scheme is decided upon will probably depend on (a) the size of the undertaking; (b) the nature of the product and the consequent degree of ease of difficulty in establishing proposed prices and (c) the amount of sophistication thought desirable.

EXPORTS

If any of the factory's output is to be exported the accountant should familiarise himself with the relevant export documentation and consult with management as to the method of payment to be arranged; credit terms; the advisability of credit; insurance and whether or not the goods are to be invoiced in sterling or foreign currency. Arranging for export contracts to be paid in sterling obviates possible profits or losses arising from fluctuations in exchange rates.

It is particularly important with export contracts to minimise the risk of differing interpretations arising from contract ambiguities or omissions, e.g. the date of sale could be the date of despatch from the factory, the date of leaving the country, or the date of receipt by the customer or his agent. Upon this date depends the date of transfer of the cost of goods sold from the company's stock account to the cost of sales account and, possibly, the date of ultimate settlement. The more precise the contract, the greater the chance of ensuring the accuracy of the company's accounts. Effective liaison must therefore be established with shipping and forwarding agents employed by the company to ensure that the correct methods of despatch of both goods and documentation are adhered to; that any special packaging provisions allowed for in the price quoted are met; and that all interested parties, including the company, receive copies of properly completed documents at the proper time.

SERVICE

If the organisation requires the employment of a field service unit, this activity should be segregated in the company's accounts. A policy decision is required at the outset when such a unit is created to establish whether a profit is to be made from the service activity, whether it is merely to cover its own expenses, or if it is to be subsidised out of the company's profits. Following this decision, the hourly rate at which service labour is to be charged out should be determined. The correct use of clearly designed forms by field service personnel will go a long way towards helping the accountant to see that the figures finish up in the right places. Such forms should show the materials used on the job and the service labour, including travelling time, should be sufficiently detailed to indicate whether or not there is a charge to be made to the customer, and in the latter case, it should be indicated whether this is due to equipment being serviced under warranty, maintenance contract or other reason. It might be helpful to allocate codes to the various categories of service calls.

The compilation of departmental expense statements is obviously of great value to management in assessing the relative efficiency of the company's operations. With regard to the service department's expense statement, it is usually considered preferable to exclude service parts sales and costs, and to show these in the same section of the accounts as the sales and costs of other products. The service department expense statement would then comprise:

(a) direct payroll costs (i.e. service engineers' wages), less
(b) service department income, plus
(c) other service expenses (suitably detailed).

SALES ADMINISTRATION

Co-ordination of the sales division's activities is exercised by sales management through the sales administration function. This involves the receipt, tabulation and follow up of reports from field salesmen, agents, sales promotion, service and distribution. The link up of any of these functions with other areas of activity within the company, e.g. research, production or finance, is usually also effected through the sales administration function. For instance, the credit stop list, which names customers taking unauthorised excessive credit and to whom no more credit is to be granted until further notice, and which is issued by accounts, would normally be routed via sales administration, whose responsibility it would be to see that the information is transmitted to interested personnel within the sales division. Amendments, adding or deleting names from the list, would naturally be dealt with in the same way.

Sales administration would also be responsible for dealing with customer enquiries and acknowledging customer orders. This entails comprehension of the meaning and implications of contract terms and abbreviations, thus minimising the risk of misunderstandings with customers and possible financial losses to the company. It is preferable for the company to have its own pre-printed official order acknowledgement forms, setting out its standard terms and conditions under which the customer's order is accepted. Such an acknowledgement should be issued to the customer immediately the order is accepted, and should the contract be of an exceptional character and not fully covered by the standard terms and conditions, management must decide if and what variations should be made. The standard points usually covered relate to:

(a) specification and quantities of goods ordered;
(b) delivery dates;
(c) basis of selling price, e.g. ex-works or delivered, indicating whether or not delivery, and even packaging, may be charged additionally;
(d) position regarding possible price increases;

(e) settlement terms;
(f) damages in case of cancellation;
(g) extent of company's liability in case of loss suffered by customer and circumstances in which applicable, and
(h) time limit for claims.

Another important sales administration function concerns the order book, often used to assist in the compilation of production forecasts. Reporting on the state of the order book enables management to assess both progress and future business, and usually takes the form of tabulating orders on hand at the start of the period, plus orders received, less orders filled (invoiced sales) and cancelled, thus giving value of orders at the period end.

It is helpful to break down the order book by product group and to analyse by anticipated delivery month.

DISTRIBUTION

Many companies adopt the concept that the manufacturing manager is responsible for the procurement of materials and the proper level of skills and facilities to ensure that the production processes are geared to the production plan, and that the finished product is thus manufactured on time.

At that point the manufacturing manager's responsibility ceases, with the sales manager being broadly responsible for the company/customer interface and, therefore, ultimately responsible for seeing that the customer receives the proper goods at the right time, supported by the appropriate documentation. Thus the sales manager's remit will take in the distribution function, involving the storage of finished goods together with their subsequent packing and despatch to customers.

In a book primarily concerned with factory accounting and administration it is not appropriate to enlarge further on this topic, important though it may be, except to make the point that the accountant should attempt to reflect the organisation in the management accounts structure to the best of his ability. His records should be maintained in such a way that departmental expense statements can be prepared easily, and supporting data should be available to facilitate, for instance, the reporting of transport costs in whatever detail may be appropriate.

5

Production control

ESTIMATING AND PLANNING

Due to the enormous diversity of manufacturing organisations and production processes, a wide measure of disagreement may be expected as to the grouping of functions under such a generic heading as production control. Once again, there is no pretence that the structure outlined here is ideal in every case. What have been included in this chapter are various functions which are linked together and which, although recognising that their appropriateness will not be universally applicable, usually lend themselves to an overall departmental type control.

The estimating and planning functions obviously take a very important place in any production organisation and in most cases these have to be carried out before any production can take place. The manufacturer engaged in complex production processes, such as computer manufacture wherein frequent changes of method may occur, or the producer working to customers' orders, possibly on a relatively short-run sub-contract basis, will usually have proportionately more estimating and planning work than occurs in a factory engaged in long-run work of a repetitive nature, such as nut and bolt manufacture. In the smaller organisation the estimating and planning functions are quite often combined and carried out by one person.

Estimating involves: (1) making an assessment of (a) the time that should be spent on each operation required to complete a particular job, (b) the rate of pay of the grade of operator, or operators, concerned and (c) an addition to cover overheads, usually according to a formula laid down by management and often quite simply expressed as a given percentage of productive labour cost; (2) determining material quantities or weights and cost, in accordance with specifications laid down where appropriate, and (3) adding the relevant company mark-up, again quite often expressed as a percentage of the total of labour, overhead and materials estimated, thus showing the profit expected from the job. According to the nature of the production circumstances, the completed estimate can give the suggested sales figure for the complete job, or it can be expressed in terms of a one-off basis, with the intention of quoting the customer the selling price of each component. The final estimate must be approved by works management before being passed to the sales department. Should sales feel that an upward or downward revision of the suggested selling price is called for, this should be settled by general management and, even though delegation may take place, the final responsibility must rest with general management.

The planning engineer, in some organisations production engineer or methods engineer, extends the estimate to incorporate more realistic times based on work study in a production atmosphere. Should this result in serious deviations from the original estimated costs, the fact must be signalled to works management immediately so that action can be taken if necessary. It might be necessary to plan the job differently or to request a price increase, and in this event the circumstance should be properly explained, e.g. whether due to increased prices of materials, wage increases, or other reasons.

METHODS ENGINEERING

Methods engineering is the modern extension of time and motion study, embracing a shift of emphasis from the ascertainment of rated times *after* the job is put on to the shop floor to establishing the best production methods *before* work. It is not proposed to write about the art of work study, but rather to attempt to give some indication of its place in an average small manufacturing unit of the kind we have in mind. Traditionally its primary aim has been to determine standard work times, usually involving incentive schemes, to reduce unit labour costs as much as possible. The stop watch has been synonymous with this activity, but it has been dispensed with by an increasing number of organisations using more sophisticated motion studies, and incentive schemes themselves, or more specifically piece work bonus schemes, have been abandoned by many larger concerns, notably car manufacturers, in favour of measured day work. It is usually accepted that any payment by results scheme should be as simple as possible, to the benefit of all concerned – the operator, the accounts staff and the management – and the bonus element should not form too large a proportion. One authority suggests not more than one-third of the normal gross wage. Production methods improvement is a broad subject and involves, apart from time and motion study, a knowledge of the machinery employed, its layout, ancillary equipment, and the materials used so that any substitution or alteration needed can be tested and appropriate recommendations made.

SCHEDULING AND SHOP LOADING

No responsible management would countenance production on a purely speculative basis though this is not to deny that calculated risks may have to be taken on occasion. Therefore the required output for a production period, commonly four or five weeks, and possibly for a number of subsequent periods, should be scheduled to form a production forecast. The required output figures may be derived from customers' delivery requirement schedules or, if building for stock, possibly from the sales forecast, modified as necessary to take account of such factors as existing stocks, machine breakdown and availability of labour. It follows that a production forecast must take into consideration (a) the volume of required production, i.e. quantities and/or weights; (b) available hours and machine time involved; (c) stock and work in progress at the commencement of the production period; (d) stock and work in progress which may be required at the end of the production period to help meet future period requirements, and (e) materials required to make the product. The shop loading function is concerned with the allocation of the work to be done according to the completion dates set, the arrangement of batch sizes and transfer deadlines between departments, where appropriate, and ensuring that management is advised in good time of unavoidable discrepancies between production requirements and productive capacity so that, for instance, where necessary any work overflow can be sub-contracted or, conversely, overtime working restricted.

PROGRESS

When the production plan for the period has been agreed, the progressing function assumes responsibility for checking performance against plan, and wherever possible anticipating likely delays and taking whatever action practicable to prevent their occurrence. Should preventive action appear to be impracticable or insufficient, progress must see that the appropriate level of management is advised if the situation warrants, in order that a decision may be made, e.g. to re-schedule or switch resources. As well as a copy of the production forecast, or schedule, the progress section will normally have a copy of the internal works order and planning sheet. The works order will give the total quantity of the part to be made, the production schedule the quantity required in the production period, and the planning sheet the sequence of operations, their time cycle, and material weights/quantities and specification. Depending on the type of production, the progress copy can be used to record quantities completed and time spent on the job, and this record can be used to highlight the fact should the times booked appear to be excessive in relation to planned times for the quantities produced.

In some cases, the times booked by operators are recorded by a time clerk on the operators' time sheets, and a suitably ruled copy works order, or production card, placed underneath so that the relevant information is carbon copied. In such a situation, the time sheets go to the wages section, of course, for payment of the employees' wages, and the production cards are actioned by the progress section.

Various planning aids are available to help works office staff carry out their duties. Among these are wall charts, which can be used by the progress section as a visual check on the current work in progress situation. Usually these take the form of a bar-chart layout with, say, the time scale in days shown horizontally and the various works order numbers shown in the left margin. Against these works order numbers, two lines are then plotted, one colour representing the scheduled production time for the job, and another representing actual time taken to date, with symbols being used to indicate certain happenings, e.g. material shortage, machine breakdown etc.

6

Supervision

INDUSTRIAL RELATIONS

Differences between individuals in intelligence, temperament, physique and ability are among the factors to be taken into account by the good supervisor. No two subordinates will be identical and the supervisor should try to avoid unwittingly creating animosity by actions which apparently ignore these differences. However, lack of effort or other forms of misconduct are different matters and, if faced with this sort of situation, the supervisor must display firmness as well as fairness. In implementing the company's policy, he must understand his subordinates and weld them into an effective team in order to get the best from them, and not fail to take appropriate action wherever necessary to create, or maintain, such a team. Whether or not the company employs a specialist training officer, the supervisor will invariably be concerned with shop-floor induction training. The best results are obtained if training is treated seriously and a formal training programme followed, and in fact the company is usually able to benefit financially if its supervisors themselves are properly trained, such as under an approved 'training within industry' scheme, as a more favourable government training grant can often be claimed.

TIME RECORDING

It is quite likely that production may suffer on occasion due to a shortage of labour or defective plant or materials. These factors apart, however, the supervisor must accept a large measure of responsibility for the work produced by his department, and for the time taken to produce that work. Management should ensure that sufficient time is allowed to carry out an effective job, before encumbering supervisors with other duties, such as machine minding or works office assistance. Regarding output, there should be an efficient, but simple, method of work measurement to confirm the quantities or weights produced by an operator, or group of operators. In the smaller factory, if the supervisor is not involved personally in checking output, it is advisable that there should be some system whereby he can signify at least conditional approval of the output level.

One way in which this could be carried out in the case, say, of work of a repetitive nature, would be for the supervisor to initial the daily quantities recorded on the work card for the operation in question. Such a set-up, of course, employs the supervisor's knowledge of the normal output level for the job and his knowledge of technical factors, such as machine capabilities and materials consumed. Whilst it is recognised that this would be by no means foolproof, the probability is that, through being made to look, the supervisor would at least be alerted to a fall in efficiency which might otherwise escape his notice. If, for instance, an operator was carrying out the same job day after day and his recorded output suddenly dropped by half, the supervisor would then be aware that this was a case for investigation. Management might have to consider instituting a system of physical checking by employing somebody from outside the section producing the output, such as a member of a production control staff. He would be responsible for seeing that the output claimed is check-counted or weighed as appropriate, and it must be decided whether such checking is to be carried out on a perpetual basis or periodically, and whether systematically or at random.

The system of time recording should include the following information:

(a) operator's clock number and/or name;
(b) part number and quantity of goods produced;
(c) works order and/or batch number, and
(d) time taken.

The monitoring and reporting of production labour times and costs is dealt with in Chapter 11.

WORK ALLOCATION

There should be a work card, or equivalent document, for each parcel of work allocated to an operator, and these could be distinctively coloured to indicate the type of work involved: e.g. production (pink), rectification (green), other non-productive (white).

Sanction must be obtained from supervision before an operator changes from one job to another, and this is especially vital before non-productive labour costs are incurred. Such costs, which are not directly recoverable from a customer, include waiting time, cleaning, meetings etc. Work should be allocated in sensible units to avoid too frequent job changes, as quite apart from the additional clerical work involved, short work spells mean loss of momentum and often disproportionate setting-up times.

As far as possible the supervisor should liaise with production control to ensure that the shop loading does not imply that the bulk of the period's production reaches its final stages towards the period end. A hold-up, for whatever reason, at a vital stage can lead to panic measures, invariably expensive, being taken in an attempt to meet the forecast period output.

Much better to smooth the output over the period, so that for instance, should key final inspection personnel be sick or on holiday in the last week, the log-jam created is reduced to a minimum and the effect on the period output is less dramatic.

MATERIALS ISSUED

The system should ensure that all materials issued to the shop floor are issued from stores against a supervisor's signature. In a short-batch production environment, it is common for a complete kit of parts to be issued sufficient to produce the batch in question.

This will vary according to circumstances, and the specialist knowledge of the supervisor and others involved will be applied in deciding exactly what is to be issued to the shop floor. Depending on the product, storage space etc., it might be decided that materials sufficient for not more than two weeks' production should be issued for any one works order at a time. An agreed scrap allowance might be added to the less expensive items, whilst items above a certain value might only be issued as required, perhaps sufficient for one day's manufacture.

The supervisor has responsibility for the materials in his area, and for seeing that they are properly used and that there are no excess stocks on the shop floor. He is also responsible for seeing that scrap is properly accounted for, and he should not allow operators to place scrapped production in a scrap bin unless covered by a scrap note issued by inspection.

MACHINES AND SAFETY

The supervisor's experience may be called upon by production control in formulating the production programme. In the event of, say, two works orders of equal priority creating a bottleneck at a certain stage, the supervisor's knowledge of the alternative machinery and equipment available, together with his assessment of the availability of the appropriate labour grades and the possibility of overtime working, could result in both orders being progressed, although not perhaps as planned in the preliminary programme.

Needless to say, any obvious errors in the production programme, or shop loading, should be pointed out to production control by the supervisor. The supervisor will know, for instance, which machines have lower output ratings, or may be aware that there are tooling problems. By taking account of setting up times, it may mean that there are just not enough available machine hours in a given production period for the job in question.

It is the supervisor's responsibility to see that the appropriate safety regulations are enforced, and any negligence on his part could render the company liable to an action under the Health and Safety at Work, etc, Act. Machines must be properly guarded, floors must be maintained in a safe condition and any oil patches cleaned or covered and protective clothing provided must be worn.

The supervisor must be alert to matters of this kind, and must not hesitate to report and take action as necessary, going through the appropriate channels to see that, for instance, maintenance department is called in at the earliest possible moment if a machine is unsafe.

7

Materials control

PURCHASE REQUISITIONING AND ORDERING

Sample forms are shown in Appendix 1. These provide the essential information required for reporting and managing the intake of materials and services.

It is the materials controller's responsibility to see that this information, or as much of it as possible, is in fact provided, and except where unavoidable, the use of such terms as 'delivery as soon as possible', or 'price to be advised', should be discouraged.

Where well-established disciplines prevail, goods are more likely to arrive on time. There will be less risk of queries being raised concerning cost allocation if the accounts code is quoted on these documents. Duplications are less likely if purchase requisitions are properly authorised and properly filed with the purchase order copy.

Financial and operational benefits rapidly result from such efficiency levels in the form of improved stock-turn ratios, and optimum use of storage space. The materials controller will normally receive from production control a copy of the production forecast detailing the materials and sub-contract content, with dates required. This should then be vetted by stock control for those items in stock to be allocated. The remaining items will need to be procured, and it may be necessary for purchasing to place enquiries with possible suppliers requesting prices and delivery dates.

Once purchase orders have been placed, production control must be advised of long-lead items so that they may decide whether or not any re-scheduling of the production plan is necessary. Of course this may not necessarily involve a complete postponement. It could be that a particular production order calls for quantities of several complex assemblies, and the fact that key parts for one or two of them are not available need not prevent the start of production of the rest.

Purchasing are responsible for seeing that delivery dates agreed with suppliers are met, and for advising production control of delays. This means employing a sensible progressing system, such as an 'up-date' file, whereby the suppliers due for 'chasing' are flagged up each day.

It is often the case that a company supplies a sub-contractor with the materials for parts to be made. Where these are supplied on a 'free-issue' basis, the materials controller must accept responsibility for seeing that the correct weights or quantities are despatched, with replacements of scrap in excess of the agreed allowance being charged to the sub-contractor. Many companies, however, dislike the 'free-issue' system, and prefer to charge sub-contractors with materials supplied, leaving it to the sub-contractor to include the materials cost with his charge for the work done when the parts are delivered.

Purchasing will normally be responsible for placing orders for items other than production requirements, such as stationery or capital equipment. In these cases, whilst the progressing function is still carried out, the 'reporting back' is usually to the originator of the purchase requisition or capital authorisation and not to production control.

The terms and conditions relating to purchase orders are of considerable importance, and are referred to in Chapter 9.

GOODS INWARDS

An infallible rule is that goods should always be booked in promptly, and the goods inwards procedures and documentation must allow for this. It is normal commercial practice for a consignor to restrict his liability by fixing a time limit on claims for damage in transit, or non-delivery, and the failure to raise timely documentation for goods received may involve the company in financial loss through being unable to claim. Also the buyer could waste time and money by urging a supplier for delivery of goods which the company had already received.

It will be clear, therefore, that prompt booking in and distribution of the relevant goods inwards documentation will go a long way towards preventing unnecessary chasing of goods already in the company, as well as seeing that interested personnel are informed as early as possible of incoming goods, thus avoiding production hold-ups and delays in passing suppliers' invoices for payment.

Usually, a copy of the purchase order is issued to the goods inwards section to assist in checking that the quantity and description of goods supplied are correct. However, the more technical the business, the more likely the need for a goods inwards inspection section, and while the broad description of the goods may be correct, the detailed description may not be. Such matters are determined by trained inspectors, possibly involving the use of special equipment, and goods supplied not to specification could be rejected if considered necessary by the goods inwards inspector.

Inspection is dealt with more fully under 'Quality control' in Chapter 8 and samples of goods inwards and rejection notes are shown in Appendix 1.

As well as goods received which have been ordered by the company, products which have been sold by the company may on occasion be returned by customers. This could be for a variety of reasons, such as incorrect goods ordered or supplied, goods defective in some form, duplicated orders etc. Such receipts are different in kind from goods received from suppliers of course, and justify the use of a separate returns inwards note. The people interested in these returns are not all the same as those involved with suppliers' goods.

STOCK CONTROL AND MATERIALS COSTING

The system of stock control adopted by a company will depend on the size of the firm, the nature and level of its stockholding and the kind of storage and administrative facilities available. Whatever the system selected, it should be as effective as circumstances permit. Frequent counting and recording of small quantities of inexpensive parts, for instance, should be avoided if possible, if this is likely to result in delays in issuing stock from stores. The cost of such uneconomic practices in terms of lost production and clerical effort may be difficult to quantify, but there is nonetheless a cost there, and as with uneconomic practices generally, such costs are best not incurred.

Careful thought must be given to the stores layout and the type of storage facility, so as to optimise the use of space. Stores equipment, e.g. steps, mechanical aids etc., also needs to be considered in stores planning, the ultimate objective being to help the storeman to find stock items easily, so that receipts and issues can be dealt with efficiently.

Stock recording, even in the smaller businesses, is not always carried out manually nowadays. There are many small computers and computer bureaux, with a wealth of stock control programmes to choose from suitable for the smaller concerns. Whether manual or computerised however, the stock control system should provide, at the very least, for the following data to be readily available:

(a) part number and description;

(b) unit of measure;

(c) stock ordered – date, order number and quantity;

(d) receipts – date, goods inwards reference and quantity;

(e) issues – date, stores requisition number and quantity, and

(f) balance of stock in hand.

In addition, many organisations use the stock record to denote minimum and maximum stock holdings, re-order levels, stock allocations and unit prices, according to the needs of the business. Stock allocations, for instance, are often used in the case of long-lead items, so that requisitions are dealt with in their proper sequence when the goods come to hand.

Pricing information is often included to facilitate the costing of materials issued and the valuation of stock. A convenient method is to use standards for pricing requisition and for stock valuation, with actual prices paid being recorded against the relevant purchase order on the stock record. There is thus a source of ready reference to actual buying prices, which can be referred to, for instance, whenever standards are updated. Standards are dealt with in Chapter 16.

Conventionally, accounting for materials normally takes the form of, firstly, allocating the cost, standard or actual, from the supplier's invoice to the appropriate stock account. Upon materials being issued from stock, the materials requisitions are costed, and then periodically summarised, possibly analysing the totals by departments or products, according to the needs of the business, with the accounting entries being to debit the appropriate work in progress account and credit stock account.

If the goods are made to order and are to be sold immediately after manufacture, the manufacturing costs, including materials used, will normally be charged to a cost of sales account, crediting work in progress. Should the goods be manufactured for stock, however, there is a further intermediate stage, charging a finished goods stock account, with cost of sales not being debited until the goods are drawn from finished goods stock and sold.

Copies of parts lists are used to compile the materials cost of each product, and the part these play in work in progress accounting is described in more detail in Chapter 13.

The degree of sophistication adopted in materials costing will vary with each organisation, and short-cut expedients may be considered justified in some instances. For example, a company may decide not to maintain separate accounts for parts, work in progress, and finished goods, and not to cost and account for materials requisitions. This would undoubtedly simplify the materials costing function, as although material requisitions would still be used for control purposes, they would involve no accounting action. Whilst the materials cost of sales totals would be arrived at as already outlined, the appropriate credit would be made to the total stock and work in progress account, with no breakdown of work in progress materials being available in the books of account.

The accountant must exercise his judgement as to just how sophisticated his systems should be, balancing the possibility of tighter control and the availability of more information to management against the time and cost of producing the information. If it is decided to adopt short cuts as described, the accountant must weigh up the pros and cons, taking into account whatever checks and controls there are in the system, and advise management of the risks, if any, in taking such a course of action. Management may decide that the risks are unacceptable and that the accounting function should be strengthened.

8

Ancillary services

QUALITY CONTROL

The quality controller may be referred to as the quality assurance manager or the chief inspector. For all practical purposes the terms are synonymous, the important thing, from the point of view of the accountant or administration manager, is an appreciation of his role and that of his department.

The control of quality starts with incoming materials and parts. Depending on the product, these may be inspected in a variety of ways, and it may even be arranged in some cases for this to be carried out at the supplier's premises. Routine ex-stock suppliers are likely to be subjected to less rigorous inspection than goods made specially to order, and may be inspected by sampling, although price considerations are important too, as incoming expensive items not properly inspected and later found to be faulty will usually result in a bigger loss to the company if used in production.

Special equipment may be necessary to check that the supplies meet the specifications laid down, and its operation may require a high degree of skill. The results of the incoming goods inspection must be recorded and this is often done by providing appropriate space on the goods inwards note, although a special form of inspection report might be required by an organisation needing to log technical data which could not be accommodated by a conventional goods inwards note.

In the event of a supplier's goods not meeting the stipulated specification, a rejection note is issued by the inspection department.

This document gives details of the goods and the reason for their rejection, together with instructions as to their disposal, such as return to supplier, scrap as agreed with supplier, rectify at supplier's cost etc.

The quality controller is also responsible for whatever checks are agreed to ensure that the goods are not being spoiled during production. This is commonly known as line inspection and is normally arranged with production management so that sensible check points are chosen. Further checks are carried out at the final inspection stage before the goods are either put into stock or desptached to a customer. The amount of inspection will vary according to the business concerned.

The cost of inspection must be balanced against the potential costs which the company could incur by the manufacture of faulty goods, with possible damage to its reputation. In certain cases, also, the production of faulty goods could expose their users, as well as the company's employees, to injury or worse. It may also be arranged that tools, jigs and gauges used in manufacture be inspected to verify tolerances etc., so as to ensure as far as possible that these are not the cause of faulty goods being produced. A register would normally be kept, examination of which would reveal when each item was last inspected, and the results.

The quality controller, then, bears a heavy responsibility and his job demands considerable expertise and judgement skills. To lessen the possible risk of undue pressure from production management to pass suspect goods, it is usually arranged that the quality controller reports directly to the managing director or general manager.

FACTORY MAINTENANCE AND GENERAL SERVICES

The maintenance department will vary in size and structure according to the needs of the firm it serves. The maintenance manager, or head, however styled, normally accepts responsibility for the maintenance and repair of company plant and buildings. As well as craftsmen such as mechanics and carpenters, this will involve unskilled people, such as cleaners.

A certain amount of maintenance and repair work could well be sub-contracted, especially if the company only carries a small maintenance section, or if the maintenance department is particularly overloaded, due to plant being overhauled during a shut-down period for instance. The maintenance manager usually acts as co-ordinator when such sub-contracting arrangements are made.

Regarding building maintenance, this can also encompass alterations to layout, when, for instance, a new production process stage is introduced and needs to be injected into an existing layout to optimise the production flow. Apart from the consequent relocation of machinery, certain fixtures such as electrical lighting and power installations may need to be re-sited or replaced, and the removal and re-erection of partitioning, or even the provision of a new exterior door etc. could be involved. The maintenance manager will be expected to organise the installation of new machinery, and it may be necessary for part of the roof to be removed to enable the machinery to be lowered into position by crane.

It is, however, with both routine and emergency plant maintenance that the maintenance manager is usually chiefly concerned. In a well-run organisation, a maintenance record is kept for each machine, so that it can be seen at a glance when the machine was last overhauled, its location, details of modifications etc. Particularly in these days of extensive computerisation and automation, preventive maintenance is of crucial importance. A machine may be set up to run a complicated sequence of operations for hours with very little attention from its operator, and if something goes wrong early in the cycle, serious and expensive damage to production could be the result. Therefore the maintenance and production staffs should work in close co-operation to minimise the risks of such eventualities. Any peculiarities in machine running, or other indications of possible trouble, should be reported as early as possible so that the problem can receive attention.

Machine down time is always a serious business, often closing a complete production line as it does, and keeping it to a minimum is high on the list of the maintenance manager's priorities.

The maintenance manager will also invariably become involved with fire and factory inspectors. He must be made aware of current legislation concerning his own sphere of activity, such as health and safety at work, and must ensure that machinery is fitted with the correct guards to prevent personal injury.

PERSONNEL AND TRAINING

These activities cover a wide and important area. In some companies both personnel managers and training officers are employed, whilst in smaller concerns the personnel function is often very much a part-time job, and could be for instance the responsibility of the company secretary.

Effective personnel management calls for a considerable degree of skill and implies a good background knowledge of personnel developments, both locally and nationally. It demands the creation of a good communications structure with managerial colleagues and the work force. Such is the importance of the role, that the personnel manager usually reports direct to the managing director.

A brief review of various aspects under five headings will provide some insight into the nature of personnel and training work.

Recruitment and dismissals

When recruiting, the personnel manager must act in as effective a manner as possible. Depending on the type of vacancy, the job could be advertised in the local or national press, will normally be registered with the Department of Employment and may also be notified to one or more employment agencies.

Before advertising, however, the personnel manager must ensure that all the essential data are available, such as age range, wage or salary, hours of work, job title, holiday entitlement and fringe benefits, if any. He may in fact be called upon to advise on one or more of these points by the manager concerned. If the job is a new one, for which there is no comparable wage or salary within the company, he may be asked to confirm local pay rates, or to advise on the rate if the job is covered by, say, a national agreement.

Job applicants should, where possible, receive the appropriate application form before attending for interview. The form should be intelligently worded, so that only relevant information is sought and the propsective interviewee is not confused.

The interview should be conducted in a systematic, but sympathetic manner, with the personnel manager handling only the introductory phase, before calling in the manager or supervisor concerned.

With regard to dismissals, whilst the actual decision is made by management or supervision, the personnel manager is normally consulted beforehand. Once the action is taken, the dismissed employee is usually passed over to the personnel manager for the dismissal procedures to be carried out and the employee informed as to how his or her final wage or salary payment is to be made, if not made on leaving. There are many reasons for dismissal, and apart from being consulted as to company policy, the personnel manager may give guidance from a legal standpoint. In the case of an employee believing himself to have been unfairly dismissed, for example, an action could be brought against the company, and the case referred to an industrial tribunal.

Contracts of employment and employee records

The personnel manager is responsible for seeing that each employee is given a contract of employment in line with legal requirements. This need not include the full set of company rules, but reference must be made to where these may be found, such as a notice board or company handbook.

Job changes within the company and changes in pay rates must be notified in writing to comply with contracts of employment legislation, as also must changes in hours or place of work etc.

Maintenance of employee records is essential to keep check of sick pay entitlement, absence etc. and periodic analysis can be helpful to management in highlighting problem areas and possible causes.

Working conditions and welfare

There are minimum standards laid down by law covering aspects such as working temperatures, periods of work between breaks and overtime restrictions for young employees. Obviously, an enlightened management will, however, strive to improve on these minimum standards and, through its personnel section, will in fact actively pursue a policy of enlightened self-interest.

This can be effected without pampering the work force. For example, to equip an assembly line with easy chairs would be more likely to result in the workers falling asleep than in any increase of efficiency. On the other hand, it would be foolish not to install correct lighting, and ensure that it is properly positioned and of the right intensity and colour to avoid glare caused by reflective parts, machinery or work surfaces. This would not only reduce eye strain, and thus possibly avoid more serious damage to the workers' eyes, but would positively ensure that output was maintained, or perhaps even improved.

Many firms have sports and social clubs which are often subsidised by the company. The personnel officer is usually the 'link-man' between the company and the social club committee, with either side able to suggest likely projects in appropriate cases. There have been many instances where local charities have benefited from the fund raising efforts arising from a variety of projects. Meals may be subsidised also, either by the provision of a canteen or luncheon vouchers. This is a way of keeping workers and attracting recruits, and the personnel officer should review the subsidy from time to time and, with the aid of the accounts department, investigate canteen prices and propose changes where appropriate.

Training and development

Apart from the advantage of a probable reduction in the training levy payable to the relevant industry training board, the consequences of a sound training and development programme are obviously beneficial. In smaller companies personnel and training are often carried out by the same person. In others there may be one or more personnel officers in addition to a training officer.

The role of the training officer does not necessarily include carrying out the actual training, for which he may or may not be qualified, but it does embrace responsibility for the training policies, and broadly the subsequent development of the workforce when trained. He must satisfy himself that the training programmes are sound and are entrusted to suitably qualified personnel.

It is recognised within industry that there will be some employees who will leave a company when trained, with offsetting benefits when a worker trained elsewhere joins the company. The attitude generally adopted is that training is for the industry as a whole with everyone deriving some gain.

Training courses may be wholly or partly 'off the job', and it is common with some apprenticeships for the apprentice to spend his first year at college, so that by the time he is working in the factory he will already have been a year in the company's employ.

Communications

This is probably the most crucial aspect of industrial relations. Resentment and grievances, real or imaginary, are often caused by inept communications, i.e. inaccurate, ambiguous, or equally as bad, the late transmission of information and instructions.

As far as possible within his power, it is the personnel officer's responsibility, to see that the company is not at fault in this respect. A common way of helping is for the personnel officer to chair periodic meetings of a joint committee comprising members of management and workers' representatives. Given the right approach by all concerned, this can be a useful medium for engendering a sense of team spirit and the avoidance of a feeling of two sides. With or without such meetings, there has to be some form of feedback, so that the intentions of the company and the feelings of the people earning their livelihoods within it are understood and taken into account.

This does not mean that a company has to reveal all. If a certain piece of information was requested of the company, and it was felt that its competitive position could be at risk if the information requested were disclosed, the company, normally through the personnel officer, should give the employees' representative concerned the reason for non-disclosure.

9

General administration and systems

FROM CUSTOMER ENQUIRY TO FINAL PAYMENT

It may be instructive to look at the various stages to be considered when a prospective customer places an enquiry, followed by a successful quotation and a satisfactory sequence of events, through to final payment for the goods delivered, or work carried out.

For this purpose let us assume that our company is prepared to manufacture to order, as opposed to, or as well as, offering 'off the shelf' delivery of proprietary products.

Stage 1   Prospective customer telephones sales manager to enquire if company has facilities to make product XYZ and is invited to send in specification.

Stage 2   Broad specification received stipulating quantity required and operating data but without drawings and tooling, these to be provided by company.

Stage 3   Sales manager advises prospect that specification received and referred to chief engineer for appraisal.

Stage 4   Chief engineer arranges for preparation of detailed specification and drawings, possibly preceded by feasibility studies being carried out by his staff, and, in conjunction with production manager, itemises tooling needed to manufacture the product.

Stage 5   Parts list prepared by drawing office and a copy of this, together with drawings, passed to production department for process layouts to be drawn up, describing method of manufacture and operations sequence.

Stage 6   Copy of parts list also given to materials control section to check stock availability of parts and for purchasing to obtain price and delivery data of parts to be bought in.

Stage 7   Estimator builds up product cost information from completed process layouts, which give direct labour grades and times, and from parts lists in co-operation with purchasing. To the direct labour and materials costs thus arrived at must be added a contingency allowance for scrap etc., plus an overhead mark-up.

The cost of tooling and of preparing drawings, plus any other engineering time, is calculated by the estimator and would normally be spread over the total product quantity involved. The final unit price would then comprise: (a) direct labour and material cost; (b) contingency allowance; (c) overhead mark-up; (d) tooling/development amortisation, and (e) standard profit percentage uplift.

Stage 8   Before the prospective customer can be quoted for the job, production management must take account of lead times and the manufacturing programme and notify the sales manager of anticipated delivery dates.

Stage 9   The completed estimate summary would be passed to the sales manager, who may wish to make allowance for market conditions and modify the proposed selling price for commercial reasons. If sufficiently important, the quotation would be reviewed with the managing director and the accountant, to take account of the impact of the possible order on the company budget and to consider the cash flow implications and the prospective customer's credit worthiness.

Stage 10 Formal quotation despatched, which will incorporate conditions of sale and indicate the period of time for which the quotation is valid.

Stage 11 Customer places order for quantity of 250 at £50 each, total order value £12,500, to be delivered in three instalments, 50 off three months from date of order, and 100 in each of months four and five.

Stage 12 Sales department checks customer's order and prepares order acknowledgement form. Any discrepancy between the order and the acknowledgement must be resolved at this point and unless the latter is modified to agree with the order, the terms and conditions stated in the acknowledgement, being the later document, take precedence in the event of dispute at a later stage, assuming that the customers had not contested the acknowledgement within a reasonable time.

Stage 13   Copy of order acknowledgement passed by sales department to production control for works order to be issued, which in turn triggers the production documentation necessary to execute the order: process layouts, job cards, materials schedule and purchase requisitions.

Stage 14   Purchasing place orders for parts to be bought in, sub-contracted operations etc., and production control enter the order on to the shop loading chart and issue the appropriate layouts and job cards in line with the work programme.

Stage 15   Work commences and accounts department summarise the weekly labour bookings and post to the relevant cost account, which could take the form of the accounts copy of the works order being suitably ruled on the back.

Accounts will also post to the cost account the materials summary total, arrived at from the materials requisitions, and the cost of sub-contract charges from purchase invoices. The key to these operations being efficiently carried out lies in sensible coding of requisitions and purchase orders.

Stage 16   Assuming satisfactory progress through production and inspection, the order is delivered as scheduled, and sales invoices raised promptly for each delivery, with accounts department issuing regular cost reports to keep management informed.

Stage 17   Customer pays accounts as they become due and final payment of £5,000 plus VAT received six months after receipt of order, i.e. one month after invoicing.

There is, of course, no golden rule which lays down that each of these stages is necessary, or indeed that there should not be more intermediate stages. Our example has been fairly straightforward and we have not considered whether any excess costs have been incurred, or cost savings made. We have assumed no production delays to throw the programme off course and, with the customer paying on time, have not needed to look at credit control and debt collection routines.

47

It will also be recognised that, according to circumstances, the time span between the first and last stages could vary enormously. In practice, too, time may not allow for all the detail work to be carried out prior to preparing the estimate summary in many cases. In such a situation, the estimator must see that sufficient contingency allowances have been built into his figures, as far as possible. Given that there will be shortcomings in almost any example, it is hoped that the chain of events illustrated offers some enlightenment as to what may happen in an everyday situation.

INTER-DEPARTMENTAL TRANSFERS

One method of inter-departmental transfers is described in Chapter 4, dealing with the transfer of goods manufactured, at predetermined prices, from production to sales department. Such a system is often employed when goods are made for stock and is a means of monitoring the relative efficiency of each department.

There are other types of inter-departmental transfers which may need to be reflected in the accounts, and the accountant must be ready to advise management of their implications. Normally, the materiality of such transfers will determine the necessity, or otherwise, of accounting action.

If there are two production departments, and each is responsible for a defined product range, a position might arise when Department A carries out work for Department B, with materials supplied by the latter. In such a case, Department B would be charged with the labour and overhead costs incurred and, provided that the labour bookings have been correctly made, this can be accommodated within the payroll reporting structure by showing the hours and labour cost involved as a transfer charge in Department A's payroll report and by supplying Department B's management with a copy. The overhead transfer charge would be shown in Department B's overhead statement as an addition to its own overheads, and in Department A's as a reduction.

The method by which such overhead transfer charges would usually be calculated would simply be to apply the standard percentage rate for the department involved. For example, if the labour charge to be levied by Department A were to be based purely on time booked, say 50 man hours at a standard hourly pay rate of £2.50, the labour cost transfer shown in the departmental payroll reports concerned would obviously be £125. If the standard overhead percentage applied to Department A labour happened to be 300 per cent, the overhead transfer charge appearing in the departmental overhead statements would be £375.

Other transfers, if material, could be necessary should normally indirect personnel, e.g. supervision, be involved temporarily as direct workers in production. Again this need not present problems as far as the accounts are concerned, if there is proper feedback of information in the form of correct booking to the job.

In another case, packaging materials charged to Department A might be transferred to Department B and an overhead cost transfer could be necessary if the amount involved was significant.

Inter-departmental accounting transfers would normally only take place where it was thought that, without such transfers, the accounts and reports would present a misleading picture because of the significance of the sums involved. Depending on the circumstances of the organisation, it could well be useful for a policy decision to be made and relayed to departmental management to minimise the possibility of inter-departmental squabbling.

STANDARD PROCEDURES

There is a sound case for someone concerned with the administration of any organisation to be given responsibility for standard procedures. Without standardisation we can easily picture a situation where Joe is clearing purchase invoices for payment and goes to the chief accountant to have them signed, whereas when Flo is doing the job she sends them to the works manager. Fred might use a proper estimating form when preparing an estimate, with Frank using the back of an envelope. When one of these employees leaves or is transferred and someone else comes in as a replacement, there is inevitably a longer learning period, coupled with a greater risk of errors, than if a written procedure is available for reference.

Although a touch of levity has been introduced to make the point, there are many real life cases of an individual being the only person to know how a particular job is done in his or her department, and it is a sobering thought that replacement in many cases can be unnecessarily costly.

As far as invoice clearance is concerned, provided that a properly authorised purchase order has been raised and that the goods inwards paperwork is in order, in practice the supplier's invoice need only to be checked, usually by the accounts department, against these documents to see that price, description and quantity of each item tallies, and there is thus no necessity for further approval to be sought. Often, the supporting documentation is stapled underneath the invoice as evidence of approval, before being posted to the appropriate accounts and filed.

A standard procedure should include the following points:

(a) an introduction to explain the purpose of the procedure;
(b) an explanation of the paperwork to be processed and its routing;
(c) clear definition of the limits of responsibility, naming job titles or departments as appropriate;
(d) where goods are concerned, clear and concise instructions as to their handling, disposal, or storage;
(e) the actions necessary to execute the procedure and the time cycle, where appropriate, and
(f) an indication of whom to refer in the event of query.

The following areas may be considered when introducing standard procedures:

(a) purchase invoice processing;
(b) costing;
(c) materials control;
(d) returns inwards;
(e) stocktaking;
(f) petty cash control;
(g) estimating;
(h) service;
(i) inspection;
(j) production control;
(k) production engineering;
(l) capital expenditure;
(m) part number allocation, and
(n) credit control.

ACCOUNTS CODE

Although it may be unfortunate, there are sometimes occasions when the accountant may need to remind his non-financial colleagues that the key purpose of keeping accounts is to enable management to assess both the performance of the business in terms of profitability, and its state of affairs, or net worth, in financial terms.

This entails the classification of all income and expenditure under appropriate headings, and the grouping of totals, or balances, to present suitably styled and meaningful statements of (a) assets and liabilities; (b) sales and other income; (c) expenses, and (d) profits and losses in summary and detailed form.

To assist in this exercise, a responsible person usually vets all source documents before they are posted to their relevant accounts. It is not normally practicable to endorse each document with the account name, which could sometimes be unwieldy, and an accounts code number is usually used instead.

It is often helpful to issue a copy of the accounts code list, or an extract, to personnel engaged in issuing purchase orders and sales invoices so that the code numbers can be entered on these in pre-arranged spaces as a guide to the accounts department, who may not otherwise be able to interpret technical terminology, when used. When such a list, or extract, is issued, in addition to the account number and name, a description or examples should be quoted as follows:

Code no.	Account name	Description
1370	Repairs and Renewals – Plant and Machinery	Includes small value Capital items (under £100)

However, the accountant must accept responsibility for the correct allocation of accounting media, and code numbers entered on the internal copies of purchase orders by the buyer, for instance, should be regarded as guidance information only, and the accountant may need to make alterations in borderline cases. An example of an accounts code structure is shown in Appendix 3.

INSURANCES

In many smaller and medium-sized factories the accountant is also the company secretary, but even if he is not, he will sometimes be involved in secretarial duties in one way or another, and insurance is a typical example.

Though he may not be involved with the negotiation and renewal of policies, or with claims, it is certain that he will be required to extract information from his accounts to enable the completion of the annual declarations mandatory with some types of policy, e.g. fire and burglary, consequential loss, product liability, employers' liability, and cash. Upon these declarations will depend the amount of premium adjustment to be made to the provisional premium charged by the insurer at the start of the insurance year.

The accountant will also find it necessary, or at least advisable, to maintain a record of insurance premiums, perhaps by means of a register. This should incorporate (a) date of charge; (b) amount; (c) policy number; (d) description of cover; (e) period of cover, and (f) amount charged to date in the management accounts.

Too often there is insufficient liaison between the accountant and the factory manager regarding the matter of insurance. It is possible, for instance, that reduced fire insurance premiums could be obtained if better 'housekeeping' disciplines were introduced, and the accountant should be prepared to seek information, when appropriate, and advise the factory manager of the results of his enquiries.

Equally, the factory manager should see that the accountant is aware of business changes which impinge on the company's insurance cover. The company may plan to 'free-issue' materials to a subcontractor for processing, and the factory manager should see that the accountant is advised, if warranted by the sums involved, so that the insurance position is cleared in the event of a fire at the subcontractor's premises for instance, just as a considerable intake of 'free-issue' materials from a customer may have insurance implications.

It is important for the company to keep the insurance aspects of its activities under review. Should the company find itself over-stocked it could also be under-insured, and when renewing policies or 'shopping around' for insurers it is wise to over, rather than under, insure. Although this cannot result in financial gain, it could prevent financial loss. In an over-insured situation, there would normally be a refund from the insurer following the annual declaration.

Insurance contracts are *uberrimae fidei*, or 'contracts of the utmost good faith', and all material facts must be disclosed, failing which the contract is liable to be revoked.

CONTRACT CONDITIONS

It is essential that the terms and conditions of all contracts, whether of insurance, employment, or the sale of goods are carefully read and understood, as non-compliance can have serious consequences.

The law relating to the sale of goods implies certain conditions and warranties, chief of which are that the seller has a legal right to sell, that the goods are of 'merchantable quality', i.e. as described in the contract, and that they are fit for the purpose specified.

Some implied conditions and warranties may be expressly cancelled in certain cases, but for this to be done the contract must be clearly written and understood by both parties.

In contrast to an insurance contract, the rule in relation to the sale of goods is *caveat emptor*, or 'let the buyer beware', although consumer protection legislation has reduced the risk to the buyer in many areas in more recent times.

For the purpose of this chapter, typical terms and conditions will be considered which may appear in contracts for the sale of goods or services, some which have also been listed in Chapter 4 in relation to order acknowledgement forms.

Goods: Quality and description to conform to specification.

Price: Must be clearly defined, e.g. ex-works or delivered, and include a statement that VAT is chargeable at rates applicable to the order. In the case of export orders FOB or CIF, and whether in sterling or foreign currency.

Payment: Whether cash with order, net monthly account etc. and whether settlement discount applicable.

Claims: Claims for loss, damage, or defects to be made in writing within a specified time limit.

Default: The action open to the aggrieved party upon default or breach of contract conditions and particularly if penalties or liquidated damages apply. There is a fine distinction between the two, with liquidated damages generally representing the estimated loss agreed by the parties, and a penalty usually a higher amount, although a court might reduce a penalty to equate with liquidated damages.

Variation: Specifying the contractual requirements should the parties agree to vary the terms. Normally a purchase order amendment would be issued.

Term: The time limit applicable to fulfilment of the order or contract and the consequences if an extension allowed, regarding prices etc.

Consequen- Normally a disclaimer is incorporated stating that com-
tial loss: pensation for any loss sustained will be limited to the cost of goods supplied and/or work done in accordance with the contract, and will not be extended to include any consequential loss suffered.

STATUTORY RETURNS

There are a number of statutory returns required of a company, and whether or not the accountant also happens to be the company secretary, he will almost certainly have an involvement with at least some of them. The Companies Act requires that all companies having share capital shall file with the Registrar of Companies a copy of the audited accounts, together with an annual return detailing:

(a) the address of the registered office;
(b) the share capital and debentures, if any, in a prescribed form;
(c) the amounts owed in respect of mortgages and charges;
(d) membership, i.e. shareholders, and
(e) directorships, together with particulars of the company secretary.

In addition to the annual return, however, there are, quite apart from such forms as VAT returns accompanying payment, government grant applications, and Industrial Training Board declarations used to determine the amount of levy payable, several returns of a purely statistical nature required by various government departments, many of which carry a legal obligation. These include monthly employment statistics, monthly sales and order book statistics, some of which may also be required in a different format quarterly, and every few years a comprehensive census of production, calling for an analysis of the company's sales, revenue and capital expenditure, employment and premises.

The number of such returns which may be required will depend on the size of the company and the industrial sector concerned, and although in the majority of cases it will not be necessary for the accountant to create special systems to cope, he will have to assess the position so that he can judge how best to collate the information required.

CUSTODY OF DOCUMENTS

It is surprising how often company management fails both to establish a clearly defined policy and to allocate responsibility to ensure its execution regarding the custody of documents.

There are three main questions which have to be answered when considering the Company's books and records:

1   In what form are they to be kept?
2   How are they to be filed and stored?
3   For how long are they to be retained?

A simple contract may be statute-barred after six years and a specialty contract, i.e. a deed or contract under seal, after twelve. Reference should be made to a suitable textbook on contract law if it is wished to examine the circumstances and conditions in which the Statute of Limitations may apply. Nonetheless many firms use these time scales as a suitable guideline to deal with question 3, but still find that they have to face up to question 1, to answer which involves categorising the records, nominally into primary and secondary groups. For example, it is highly unlikely that a company would wish to preserve petty cash vouchers and purchase requisitions for the same length of time as invoices. It is, therefore, likely that some records of lesser importance may not be kept for six years, while others, such as year-end accounts, contracts of employment etc., may be retained indefinitely.

The form in which records are to be kept need not be as original documents. They could be microfilmed and the originals destroyed, or it could be decided to microfilm and to retain the originals in a separate location as security in case of fire. It could further be decided that only selected groups of primary records of an archival nature are to be microfilmed, and that the current and most recent year's records are to be kept in 'hard copy' form.

If the company has its own in-house computer a decision will have to be made as to the length of time various classes of print-outs are to be kept. It may not be necessary to keep hard copy data for as long a period as would be the case if the company did not possess the master tapes or disks.

Security is a vital factor, as regards both loss through fire or theft and the confidentiality aspect, and serious thought needs to be given to question 2. There is a wide range of safes, cupboards and cabinets to choose from for the purpose of storing the more sensitive documentation and as far as the remaining records, probably the bulk, are concerned the main considerations will usually be accessability and space available with the choice of files, racks, trays etc. being made accordingly.

To facilitate the formulation of policy by management it is a good idea for each department to draw up a list of its own records, suitably categorised, with a recommended retention programme. This will often reveal an element of duplication of certain records, which may be unavoidable or at least acceptable for current usage, but which may be painlessly eliminated at the archive stage.

Busy management will invariably be fully occupied in running the business, but a systematic approach to rationalising paperwork storage and retention with due attention to efficient data retrieval, without 'going overboard', can be rewarding and may save countless hours of unnecessary searching.

10

Overhead allocation and control

PAYROLL COSTS

All payroll costs which cannot be apportioned conveniently to the manufacture of products with any degree of accuracy should be treated as overheads. This applies to the employers' costs of national insurance, pension contributions and other employee benefits, in addition to gross pay.

Direct labour comprises those payroll costs directly chargeable to jobs, normally derived from an analysis of time records. Labour costing is dealt with in Chapter 11.

Time spent by normally direct operators on non-productive work, such as machine cleaning, stocktaking etc., is costed as overheads, as is usually overtime premium pay, with only the basic rate content of the overtime hours being charged to the job.

It is, nonetheless, helpful if overhead payroll costs are allocated to appropriate cost centres, usually departmental, and apart from being useful in the construction of accounts and reports, this can be of considerable assistance to the estimator in arriving at the recommended selling prices of the company's products.

The consideration of seemingly academic issues such as cost allocation should obviously never be allowed to divert management's attention from the paramount task of ensuring that all employees, direct and indirect, are productively employed.

Of course the question 'Is this cost necessary?' must be asked of all forms of expenditure. This is particularly so in the case of payroll costs, however, as every additional employee causes increases in other overhead costs, such as welfare expenses, consumable supplies etc.

There are certain levels below which numbers employed must not be allowed to fall if the company's viability and potential are not to be threatened. It is a matter of judgement as to what these levels are at any given time and in any given area, and this demands the exercise of considerable management skill and discretion.

SERVICES AND SUPPLIES

Every commitment to spend money requires prior examination and approval, and the system under which this is carried out needs to be sensibly organised to avoid needless bureaucracy. There should be less detailed examination of relatively minor items of expenditure, or of known standard charges, such as equipment hire, but closer scrutiny is necessary where higher costs are concerned.

Direct materials costs are often rigidly controlled, perhaps because the buyer is working from a materials requirement plan built up from parts lists, or possibly because those concerned within the company may be aware that there is a more or less standard weekly or monthly intake of a particular product.

There is not always, however, the same control of overhead expenditure, and one can often find cases of indiscriminate use of telephones, over-stocking of packing materials and extravagant spending on stationery.

Sensible budgetary control will help to ensure that overheads do not get out of hand and it is a good idea to arrange for departmental managers and section heads to have authority to approve defined types of expenditure up to pre-established limits. As with most arrangements, such a scheme would need to be reviewed periodically to ensure that the scope and limits of expenditure were still appropriate to the needs of the business.

The various types of services and supplies often included in overhead statements can be seen in Appendix 4, which illustrates a typical management accounts layout.

DEPARTMENTAL COST CONTROL

The allocation, collection and reporting of costs associated with operating a department is an important function of the accountant. Departmental and general management are vitally interested in departmental operating costs, and separate cost statements for each department will usually be called for.

It follows that, for these objectives to be attained, careful analysis of payroll and other expenditure is necessary, and payrolls, ledgers and subsidiary books should be organised in such a way as to accommodate this requirement. Books and accounts kept manually could, for instance, be ruled to allow for columns for each department so that entries can be made direct into the appropriate departmental column, thus saving the need for further analysis after the original entry.

Whether the books are kept manually or not, it is a sound practice to employ a code of accounts, so that the relevant code number can be entered in an appropriate place on source documents. A sample accounts code is shown in Appendix 3, and copies of the selected accounts code list should be issued to all personnel with responsibility for, or an interest in, its operation. Some parts of the code may not be relevant to certain personnel in some cases. An example of this would be the buyer not needing to know the various code numbers of the capital and reserves section, and in such situations extracts only need be issued.

Departmental costs can have a direct bearing on pricing, profitability, cash flow, and even employment policies and it is important that there should be proper communication between the accountant and departmental management. This involves explaining and listening, learning and teaching, which cannot fail to make all concerned better at their jobs.

FINANCE, ADMINISTRATION AND ESTABLISHMENT CHARGES

As always in this book what follows is meant to be illustrative, rather than definitive, and many firms may not accept this particular grouping, preferring to use variants of their own.

Included under this heading are those overhead items which cannot be directly allocated as manufacturing or selling expenses. This is, of course, a broad statement and in practice there could be many more overhead cost centres, particularly in the more complex organisation.

A possible structure could comprise:

(a) research and development;
(b) manufacturing;
(c) selling;
(d) distribution, and
(e) service.

Circumstances may dictate further subdivision, e.g. manufacturing – product A, manufacturing – product B etc. Finance, administration and establishment costs would be in addition to these overheads, and could themselves be combined in one overhead statement or subdivided.

In this example it is assumed that premises are shared by all departments. If this were not so and each department had its own building, then establishment, or occupancy, charges could be allocated direct to the departments concerned.

Examples of the type of expense appearing under each of these subheadings might be:

Finance:	audit
	bad debts
	bank charges and interest.
Administration:	general office salaries
	legal and professional fees
	postage
	telephones.
Establishment:	building repairs
	leasehold amortisation
	rent and rates.

The principle should be that these overheads, as with all others, should be compiled, as far as possible, in such a way that an overhead statement can be presented to the director or manager responsible for the control of the items appearing in it.

ALLOCATION OF MANUFACTURING AND SELLING OVERHEADS

Working on the basis that the cost of selling, distributing and servicing products is incurred after their manufacture, non-manufacturing costs should be excluded from the overhead content of stocks and work in progress.

Using imaginary figures to illustrate the point, assume the following data:

	£
Direct labour	6,000
Manufacturing overheads	20,000
Selling and distribution	5,000
Finance, administration and establishment	8,000

If the agreed apportionment of finance, administration and establishment costs worked out at, say, 75 per cent to manufacturing and 25 per cent to selling and distribution, the figures could be re-stated thus:

	£
Direct labour	6,000
Manufacturing overheads (including FAE)	26,000
Selling and distribution (including FAE)	7,000

Of the total overheads of £33,000, note that £26,000, ignoring variances, would rank for inclusion in stocks and work in progress. In this example, if the overhead allocation was based on direct labour cost, the overhead rate would be expressed as 433 per cent (i.e. 26,000 ÷ 6,000).

Whilst it is necessary to avoid becoming unduly bureaucratic, the essential point is that the apportionment of finance, administration and establishment costs, to manufacturing and selling overheads should be seen to be fair. Therefore the accountant should set out clearly the basis adopted. He may prepare a list of expense items and, with due consultation, show alongside each the estimated apportionment, e.g.:

	Manufacturing	Selling and distribution
	%	%
Accounts department costs	75	25
Rent and rates	80	20
Telephone	60	40

An agreed base period, perhaps from the budget or from the previous six months' accounts, could be taken, and values slotted under each heading, giving totals which could be expressed in percentage terms and applied to actual costs incurred from that point on. The apportionment basis could be reviewed periodically, perhaps twice yearly, to see whether the percentages chosen were still valid.

11

Payroll accounting

PAYROLL ORGANISATION AND PAYMENT METHODS

The organisation of the payroll function is, arguably, the most important administrative facet in any enterprise employing people, and in an industrial concern its importance is crucial.

Without an efficiently run payroll unit, management has less chance of timely and reliable payroll data, e.g. departmental costs, overtime worked, holiday pay, cost of employee benefits etc. In an extreme case this could result in incorrect or late payment of wages and salaries with an inevitably detrimental effect on industrial relations.

Whether or not wholly manual systems are employed, the payroll should be organised in such a way that:

(a) there are separate payrolls for different pay periods, e.g. monthly and weekly;
(b) there are separate payrolls for separate locations, e.g. branches, if justified by numbers employed;
(c) departmental costs can be analysed;
(d) the cost of additional payments, e.g. overtime, bonuses, holiday pay, can be seen;
(e) tax and other deductions can be analysed correctly;
(f) employers' cost of national insurance can be totalled, and
(g) individuals and total net pay is apparent, so that wages cheques can be correctly drawn, and pay packets made up.

Organisation of the payroll function must include such elementary, but vital, matters as security arrangements, regarding both cash and confidentiality of pay records, and ensuring that cheque signatories are available at appropriate times. Attention to details of this sort will help to ensure payroll efficiency.

Methods of payment – cash, cheque or credit transfer – may be influenced by numbers employed, location of pay centres, frequency of payment, and degree of complexity in calculating pay and deductions, such as production bonuses, overtime, holiday pay entitlement, laundry charges etc.

From the security aspect and the work angle there are distinct advantages in payment by credit transfer, and this method usually predominates where pay is remitted monthly. The bulk of the workforce in most factories, however, is paid at more frequent intervals, usually weekly, and due to the time taken to reach the recipient's bank account, up to a week, payment by credit transfer is not normally acceptable in such cases.

Even so payment does not necessarily have to be in cash. It is worthwhile striving to obtain agreement to payment by cheque, and if the timetable permits and the accountant can suitably organise the payroll function and make arrangements with the appropriate managers and workers' representatives, there are dividends to be gained. The task of making up wage packets is eliminated, as is the risk of wages theft and personal injury to staff. There would also be a clear benefit in the form of lower insurance premiums. It can be seen that this is an objective well worth aiming for.

MANAGEMENT REPORTING

Regular reporting of relevant data to the appropriate levels of management can only be beneficial. The accountant must, however, keep the keywords in the forefront of his mind, i.e. regular, relevant and appropriate.

Regular reports will ensure that management is kept informed and, more important, is in a position to take timely action when necessary. That the data reported should be relevant may appear obvious, but it is all too easy for irrelevant data to creep in if the main purpose of the reports generated is lost sight of, probably leading to valuable management time being wasted sifting and digesting the information required, and possibly causing confusion and even apathy.

The question of time wasting and confusion may also occur if a report is directed to an inappropriate management level, e.g. where a detailed weekly departmental payroll report is sent to the supervisor of another department, or to the company chairman. In the first case the supervisor is only concerned with the detail relating to his own department, and in the second the chairman does not, or should not, need the information in such detail anyway.

It is useful to issue a daily list of non-productive labour bookings, as well as a weekly payroll report. Sample reports are shown in Appendix 2, and a summary of the information it should contain is as follows:

(a) name of department/section;
(b) non-productive hours spent by direct workers, broken down into main headings, e.g. waiting time, maintenance and cleaning of machines and tools, stock-taking, meetings etc. (these could be denoted by agreed code numbers);
(c) date.

This report could consist of a single sheet of paper covering one week's bookings, with columns for each day, each column to be initialled by the appropriate production supervisor daily.

The weekly totals could then be included in the more comprehensive weekly payroll report which would comprise the following information:

(a) an analysis of direct, i.e. productive, hours worked by
    (i)   department or section, and
    (ii) product group;
(b) an analysis of indirect hours booked by nominally direct workers, as outlined above;
(c) cost of productive labour sections analysed by
    (i)   direct work, broken down into product groups,
    (ii) total indirect labour cost,
    (iii) overtime premium plus, in the case of standard costing,
    (iv) labour variances, which can be further analysed between rate and efficiency variances;
(d) cost of indirect labour sections analysed by department or section, e.g. inspection and production control;
(e) employers' cost of national insurance;
(f) budget totals for each of the above items;
(g) employers' gross payroll cost;

(h) payroll deductions itemised, i.e. PAYE, national insurance etc., and

(i) net payroll cost.

LABOUR COSTING AND THE USE OF STANDARDS

It would be wonderful if, at the end of every accounting period, stock and work in progress did not exist. Costing would be simplicity itself and the accountant would find that managers would have less scope to contest his figures.

The cost of all materials purchased and work done, even if incomplete, would be charged out, and the profit (or loss?) could be worked out in five minutes flat on the back of the proverbial envelope.

In the real world we know that we have to get to grips with stock, and even worse, work in progress. Assuming a sensible approach, however, the problems need not be insurmountable.

It is usually costly and impractical to check stock physically and work in progress at the end of each interim accounting period, but we can nevertheless so organise affairs that, with the aid of even rudimentary systems, reasoned estimates can be arrived at.

In preparing the quotation for submission to the customer, the estimator would have reached certain conclusions as to

(a) the categories of labour to be employed;
(b) the time to be taken by each to produce a given quantity of the article in question;
(c) rates of pay, and
(d) the cost of the labour content of the job.

The estimator would also include such contingencies for inflation etc. as would appear prudent.

In the event of the customer responding to the quotation by placing an order, the estimate would normally be refined by the planning engineer, as outlined in Chapter 5, and realistic production times set for each operation and sub-assemblies, if any. Allowances for fatigue and personal needs would be included in the times set.

From his copy of the planning card the accountant can now devise labour standards, and by comparing times taken with planned, or standard, times his costing section can monitor efficiency variances, which must be regularly signalled to shopfloor supervision for action where necessary. As shown in the example at the end of this chapter, these efficiency variances can be segregated so that they do not find their way into the work in progress control account.

Calculation of the labour standard will have involved arriving at a pay rate, as well as work times. Whilst variations from standard work times are dealt with as efficiency variances, variations in rates of pay are likewise segregated from work in progress and are reported as rate variances.

Taking the case of an instrument-making section, which included, say, a small number of senior operators at higher rates of pay, and a larger number of average operators at lower pay rates, an overall average pay rate for the section could be arrived at. This rate could then be adjusted to allow for certain anticipated factors, e.g. pay increases, and used as a standard for the ensuing six months. If at any point during this time there happened to be a preponderance of one type of labour, due to the influences of holidays, overtime etc., this would naturally give rise to a rate variance.

Thus, by regulating labour input to work in progress, and excluding labour variances, we have ensured that labour input is measured at standard. This is precisely how our labour cost of sales is arrived at and as this forms our work in progress labour output, it follows that, provided the opening balance has also been computed at standard, the closing balance of work in progress labour represents its prima facie standard value.

In fact, there are certain complications which invariably arise, usually of a minor nature, and some allowance must be made for these, even if only in the form of a round sum provision, based on past experience, debiting excess costs and crediting work in progress.

For example, an article may be scrapped when 90 per cent complete, and the labour invested up to that point, which has been included in work in progress input, must now be taken out of work in progress. The use of a scrap reporting system employing a scrap note as shown in Appendix 1 will help in dealing with this problem.

Again, a machine may break down and to maintain production a slower machine may be used. If the standard has been based on the faster machine time, the charge to work in progress will be overstated unless steps are taken to prevent this. The excess time involved should be properly authorised and written off, perhaps by the use of a form similar to the temporary operation note, also shown in Appendix 1.

Example: Assume (a) article contains standard of 20 direct labour hours at £2.50 per hour;

(b) average pay rate for period £2.35;
(c) times booked = 200 hours;
(d) standard time for work done = 180 hours;
(e) 5 units sold.

The accounting entries could be made in the following form:

	Debit £	Credit £
Work in progress labour:		
180 hours at £2.50	450	
100 hours at £2.50		250
Labour rate variance:		
200 hours at £0.15		30
Labour efficiency variance:		
20 hours at £2.50	50	
Cost of sales labour:		
100 hours at £2.50	250	
Payroll control:		
200 hours at £2.35		470
	750	750

Note: Payroll report would include (a) work in progress   450.00
                              (b) efficiency variance   50.00

                                                       500.00
                              (c) rate variance   30.00

Actual payroll cost, i.e. 200 hours at £2.35   £470.00

A different system may be appropriate if there is a high degree of automation and output is determined primarily by the machine, with the operator playing a relatively minor role. Machine hour accounting is touched upon in Chapter 13.

12

Controlling liquidity

CASH FORECASTING AND THE USE OF CASH FLOW STATEMENTS

A distinction must be drawn between short- and long-term cash projections. Companies sometimes reflect this by the use of differently styled statements which give recognition to the fact that, whereas a budget is commonly drawn up to cover a period of twelve months, a shorter-term management tool can be helpful for a period of, say, one month ahead, particularly so in the later months of the financial year when, for a variety of reasons, the year-to-date figures may differ considerably from plan.

A typical cash forecast would be prepared at the end of the month and would attempt to predict the bank balance at the end of the following month by itemising likely receipts and payments. Actual receipts and payments could be reported on a weekly basis and compared with the forecast, giving an indication of progress being made against the prediction.

It is possible that the preliminary forecast may be unacceptable, e.g. if it revealed that an overdraft limit would be exceeded. In such a case an examination of the situation would probably result in the preliminary forecast not being issued, and a re-drawn forecast prepared, perhaps reducing the amount allocated for the payment of creditors' accounts.

The accountant would, of course, apprise management of the serious situation necessitating such action, as there would be an obvious danger of the withdrawal of credit by suppliers and, perhaps, legal action being taken against the company, unless it were possible to negotiate a temporary extension of normal credit terms. Other possibilities might be to arrange an increase in the overdraft limit or to step up the debt collection activity.

The whole point of the exercise is to create an awareness of the probable future position and once the forecast is issued the weekly cash reporting would serve to generate action to ensure that it would be met. An example of a simple cash forecast is given in Appendix 2.

In the longer term it is useful to include a cash flow statement in the budget, giving an indication of the monthly peaks and troughs, from which it would be seen, among other things, whether there was a need to reschedule the purchasing programme to avoid an excessive stock build.

Such statements are compiled simply by listing changes in balance sheet values and the inclusion of a cash flow statement in the management accounts enables monthly comparisons to be made with budget, highlighting problem areas and serving as a check that the company is on course in the longer term.

Cases of profitable businesses with cash problems are legion, and with the aid of the cash flow statement the accountant must be prepared to draw to the attention of his colleagues in management the consequences of spending more on capital investment and building stocks than the amount of cash generated from profits.

Typically, the statement will show the budgeted and actual figures, both for the current month and year-to-date, relating to opening and closing stocks, debtors, and creditors and thus the movement of each, pre-tax profit (or less), depreciation, capital expenditure, disposal of fixed assets, taxation, and exceptional items such as long-term loans, dividend payments etc. A cash flow statement is included in the set of sample management accounts under Appendix 4.

Under the auspices of the leading accountancy bodies, 1972 saw the introduction of accounting standards. The accountant should have access to these statements, known as Statements of Standard Accounting Practice, and their intention is to describe approved accounting methods in order that accountants may provide an improved service to users of financial statements generally.

SSAP 10, 'Statements of Source and Application of Funds', was issued in 1975 with the declared objective of including such statements as part of the audited accounts. The standard allows some flexibility in the presentation of the required information, but a survey of various audited accounts has shown that there is a general preference for the format illustrated in the appendix to the standard, at least for the purposes of the statutory accounts, although this format is also used in the monthly management accounts in many cases.

The accountant must, therefore, reflect on the needs of managers and shareholders and whether they can best be served by identically prepared statements, or whether, indeed, different forms of presentation would be preferable.

A funds statement, similar to the appendix to SSAP 10, is shown below and it can be seen that, although both forms essentially reflect changes in balance sheet values, the cash flow statement illustrated in Appendix 4 contains important differences in presentation, chief of which is the emphasis given to the cash movement, shown as the statement total in preference to profit plus depreciation.

Source and application of funds

		£000
Source:		
Net profit before tax		98.7
Depreciation		21.2
		119.9
Application:		
Increase in fixed assets		47.5
Increase in cash		25.7
Increase in debtors		16.9
Decrease in current liabilities		11.1
Increase in stocks		32.0
Decrease in amounts owed to affiliates		2.0
		135.2
Decrease in prepayments	2.1	
Increase in deferred income and liabilities	13.2	
		15.3
		119.9

SCHEDULING COMMITMENTS

An aid to the accountant in the compilation of cash flow budgets and medium-term cash forecasts is sometimes provided by purchasing department supplying a monthly summary of committed purchases. This gives a longer-term picture than could be obtained by 'ageing' the purchase ledger balances and, in essence, consists of analysing outstanding purchase orders by anticipated month of delivery, thus arriving at committed costs for up to, say, six months or more ahead.

To avoid a complete rescheduling exercise every month-end, a register would normally be kept into which the purchase order costs are entered as orders are raised with the register containing analysis columns, one for each month of delivery.

It would probably be necessary for adjustments to be made, usually arising from purchase order amendments, and due to one or more of the following:

(a) cancellations;
(b) price changes;
(c) goods returned;
(d) short/over deliveries;
(e) rescheduled deliveries, and
(f) changes of items/quantities ordered.

To complete the exercise at month-end closing, the accountant would need to compare the purchases total with the forecast commitment and enquire into any sizeable discrepancies.

Scheduled commitments are frequently also broken down into product groups.

CREDIT CONTROL

Conventionally, credit control is taken to be synonymous with the control of debtors. It does not, however, relate strictly to debtors alone.

An important element in the control of a company's resources concerns credit allowed by suppliers and the term 'credit management' is sometimes used to embrace both aspects, credit given and credit received.

Every encouragement should be given to the obtaining of favourable credit terms or the negotiation of settlement discounts, while the granting of cash discounts to customers, which frequently work out to be more expensive than the cost of bank interest, should be avoided if possible.

There is little point in paying creditors' accounts before they become due and the system should not allow this to happen by accident. Although unlikely to be as potentially injurious to a firm's operations as late payment, the early payment of accounts for no apparent reason is an unwise and often costly diminution of cash resources.

Turning to debtors, effective credit control implies status checks on prospective customers before credit is granted. Naturally, some discretion will be required and the approach therefore will inevitably be selective. It would not be sensible to delay processing an order, for instance, if the new prospect happened to be a government agency or ICI and there could be other cases, particularly if the amounts involved were not large, when it might be decided to dispense with a status check, such as an order received from a large firm, perhaps in the same locality, with which the company had had no previous dealings.

In other cases, one or more of several methods would be used to obtain satisfaction of creditworthiness. Typical of these are (a) trade references, (b) bank reference, (c) report from a recognised status enquiry agency, and (d) guarantee from a third party, possibly a parent company. In the light of the replies received, a decision can then be made as to the amount of credit that can be allowed and sales management advised accordingly.

Once credit has been allowed and an account becomes overdue or a credit limit is exceeded, action must be taken to correct the situation. It can be worthwhile employing a standard procedure to cover credit control, with copies being issued to accounts personnel involved and interested members of management, including sales.

The procedure should include an introduction to explain its purpose, and should specify the method of dealing with new accounts, and the stages to be followed in the recovery of accounts when credit terms are exceeded, perhaps using standard letters, naming the department or person by function responsible for action at each stage.

It need not be necessary to adhere too rigidly to the procedure and some flexibility should be preserved to allow the odd telephone call, or personal contact from sales department, provided that this is likely to assist, rather than delay, recovery.

Should the point be reached when credit has to be withdrawn and legal action threatened, it is important that the appropriate level of management is notified, so that the despatch of further supplies may be prevented.

When legal action is necessary, it could be advantageous to look at the possibility of using the county court procedures for the smaller debts, currently under £2,000, and it may or may not be desirable to use the company's solicitors, depending upon the circumstances of the case.

The recovery of debts from overseas customers is sometimes a more hazardous operation, but not necessarily so.

There are various services available, such as those of the Export Credits Guarantee Department and specialist insurers prepared to cover bad debt risks. In many cases, however, the company will arrange for the debtor to supply a documentary letter of credit and useful advice and literature can be obtained from the company's bankers regarding this and other methods of payment by overseas customers.

Arrangements whereby the customer does not receive possession of the documents entitling him to the goods until payment, often by sight draft, has been made to the bank's agent are usually satisfactory in cases where this is an available option.

EXPENDITURE CONTROL

Bearing in mind that liquid capital is composed of short- to medium-term debtors, creditors and bank balances, it will be appreciated that the control of expenditure has a direct, often immediate, bearing on liquidity.

It is essential that a sound system of expenditure approval be maintained as a check on possible waste and extravagance. Even when the expenditure is agreed to be necessary, it is invariably worth asking if it can be deferred when significant sums are involved, unless the expenditure request makes the timing clear.

Each case needs to be judged on its merits and if an argument is advanced in favour of buying in, say, nine months' stock of a particular part to avoid a price increase from supplier, the estimated saving must be weighed against the cost of holding the excess stock.

Expenditure control, therefore, starts before the expenditure is initiated, let alone incurred, and the application of sensible procedures will ensure that, once sanctioned, the timing and cost are properly monitored, with descrepancies being observed and reported.

Purchase requisitions and capital authorisations dealing with their separate aspects of expenditure control are referred to in Chapters 7 and 15 respectively, but it is no less important that other types of expenditure, ranging from the sanction of overseas travel to petty cash expenses, and from the authorisation of overtime working to additional staff recruitment, are also effectively supervised.

Whatever the paperwork format, the signature of a responsible designated official should be obtained. Taking petty cash as an example, the cashier should be given a list of managers and supervisers, by job title, with authority to sign for expenses incurred by their departments, and any delegation of this authority should be in writing. It is a good idea to fix an upper limit for petty cash expenditure, with amounts over, say, £50 being paid by cheque, and cash sums in excess of this being paid only if specially agreed as essential. This avoids the need to carry a larger cash float than would otherwise be the case, and the operation of the imprest system, whereby only the cash expended from the float is reimbursed, can also help prevent too large a petty cash balance being maintained.

In controlling the bank balance, the accountant should be prepared to contribute in the formulation of policy decisions, particularly when capital expenditure is being reviewed. Depending on circumstances, it could be decided that the cost of exceptional capital expenditure should not be charged to the bank current account as incurred, but spread over a longer period.

This could be by the arrangement of hire-purchase facilities for instance, or by the negotiation of a long-term loan, possibly from the bank. In other cases it might be decided to lease, rather than buy, the equipment concerned.

13

Stocks and work in progress – accounting and control

STOCKTAKING DISCIPLINES

In most types of manufacturing businesses it is necessary to carry out a physical check of stocks and work in progress in some shape or form. This would certainly be so in the case of a manufacturer producing a standard range of units, such as electric motors, but even a single item major project contractor involved, for instance, in the electrical power and lighting installation of a warehouse, will usually need to check stocks periodically, whether to satisfy its own management, the auditors, or the customer as to the state of progress of the project at any particular time.

Given that a physical check is to be carried out, methodical planning and execution are essential. The accountant should issue clear and concise instructions in writing to the appropriate levels of management and supervision engaged in the exercise, and this should be preceded by a study of the problem and discussion with management.

Some points for consideration in compiling the stocktaking instructions are as follows:

1   Preparation – The areas involved should be made as tidy as possible and the materials and parts to be checked should be clearly identifiable. Any equipment necessary for an efficient stock check, such as scales, pallet trucks etc., should be available for the commencement of stocktaking.
2   Type of check – It may be decided that it would be too costly and time consuming to carry out a full check. The extent to

which this can safely be dispensed with will depend partly on the system of stock recording in force and partly on the likely total value of the stock concerned. Possibly items of low value would only be spot-checked or perhaps not counted at all.

To avoid the annual or twice-yearly chore of stocktaking, some firms employ a perpetual inventory system whereby various categories of stock are counted in rotation so that all items to be counted are counted at least once during the course of the year. Perpetual inventory implies the correction of stock records to adjust for discrepancies, and periodic lists of discrepancies to the accounts department so that the appropriate accounting adjustments may be made.

3   Documentation – Pre-numbered stock sheets and work-in-progress dockets should be used, and these should provide for all relevant information to be included, e.g. part number, description, unit of measure, quantity and, in the case of work-in-progress dockets, the completion stage. There should also be space allowed for unit and total cost of each item, broken down as necessary between materials and labour. If possible, stocktaking documentation should be pre-printed, leaving only quantities and values to be recorded.

4   Cut-off – Goods received prior to taking stock should be booked in as such to ensure as far as possible that the relevant suppliers' invoice is entered in the books in the correct period and not in an accounting period subsequent to the stock-take date. Likewise, goods received after taking stock should be booked in and the purchase invoice recorded in an accounting period following stocktaking.

To avoid goods being included both as sales and stock, any returns from customers must be credited before the date of taking stock. It is a good idea for the last goods inwards note number before stocktaking to be advised to the accounts department and for subsequent receipts to be recorded at least two clear days later to reduce the risk of confusion. These comments apply similarly to outwards advice notes.

5   Responsibilities – The stocktaking instructions should unambiguously set out the specific areas for which particular members of the management and supervision team involved are to be held responsible.

STOCK AT THIRD PARTIES AND FREE ISSUE MATERIALS

It is important when stocktaking to see that any company property at third parties, such as sub-contractors holding goods for processing or goods on loan to customers, is included in the stock count, and suitable arrangements should be put in hand to cover this point, preferably entailing confirmation from the holder of the goods.

Similarly, goods on the company's premises which are the property of third parties should not be included in stocktaking and should be segregated in a bonded store.

Although items of both sorts are not normally paid for except in the case of loss or damage, there is obviously a need for an effective control system on a day-to-day basis, quite apart from stocktaking requirements. For this reason, many firms arrange for 'free-issue' stock, both in and out, to be charged on a memorandum basis and credited on return. The memoranda debtor and creditor accounts provide a useful means of determining the total sums involved at a given point in time.

VALUATION BASES

Leaving aside the inflation accounting debate, it is a tried and tested accounting convention that stock should be valued at the lower of cost or market value. This still leaves considerable scope for variations in treatment between the pessimists and the optimists, due to differences in policy and definition of terms. Therefore, the valuation of stock is highly unlikely to be 100 per cent objective, as an element of opinion is almost certain to enter the equation somewhere along the line. This is particularly true of 'market value' for instance, the determination of which will usually depend on whether the director or manager concerned is an optimist or a pessimist.

Regarding policy, the accountant should be prepared to help formulate the criteria for establishing the stock provisions necessary to write down obsolete and slow moving stocks. Obsolete stock, by definition, has no value to the company and should be disposed of in the most effective manner and written out of the books. If, for some reason, it is decided to keep obsolete goods in stock, they should be segregated from the main stock, if possible, and the stock value of the obsolete items should be credited to a stock provision account and debited to a stock adjustment account. Similar accounting treatment is recommended in dealing with slow moving items, except that in this case, only part of the stock value is written off. This could be done in a sophisticated manner by, for instance, writing off varying percentages according to the age of the items concerned, or a simpler procedure could be adopted, such as writing off 50 per cent of the cost of all slow moving goods.

Cost is open to a number of interpretations, as far as stock and work in progress are concerned, and one or more of the following bases is usually followed:

1 First in, first out (FIFO), assumes that the older stock items are consumed first, with stock valuation being determined by the cost of the later items.

2 Last in, first out (LIFO) is the opposite of FIFO, and assumes that stock is valued at the cost of the older items.

3 Average cost. This involves calculating a new average cost for each consignment received if the price is different from the previous average.

4 Base stock value is used sometimes in situations where massive stock fluctuations do not occur, with no change being made in the, normally conservative, stock valuation for the particular range of stocks involved.

5 Standard cost. Stocks in this case are valued at predetermined standard prices. The principles of standard costing are outlined in Chapter 16.

CLASSIFICATION OF STOCKS

It is usually convenient to keep separate records of the following different types of stock:

(a) raw materials – such as metal stocks and sheets of material from which a product is made;

(b) parts – components purchased for incorporation into products;

(c) work in progress – incomplete products in course of manufacture;

(d) finished goods – completed products in stock pending despatch and sale to customers;

(e) loose tools – smaller items of equipment used in manufacturing company products, such as hand drills, screwdrivers etc., and

(f) consumables – oils, greases, cleaning materials etc.

Further analysis may be required in some of these classifications, depending on circumstances. This could possibly be by product ranges, in the case of a company making more than one product range, and/or by department, such as machine shop, assembly, and plating shop.

PROCESS LAYOUTS, PARTS LISTS AND WORK IN PROGRESS ACCOUNTING

Process layouts and parts lists play a vital part in a well organised system of work in progress accounting. The process layout, also known as a planning card, describes the work to be done, and the materials and equipment to be used, to manufacture a certain component or build an assembly.

The form heading will usually contain the following information: part number and description; used on or required for; materials or parts to be used (this space could be endorsed 'see parts list attached' where many parts were involved, e.g. an assembly); quantity to be made.

Beneath the heading there would normally be columns in which would be entered: operation number and cost centre (e.g. department); operation description; tooling or machinery to be used; set-up time; output rate required (variously termed a standard, target, or allowed time and expressed either as a quantity per fixed time, e.g. an hour, or as a time per fixed quantity, e.g. minutes per 100).

The parts list for a given job will normally contain the following information: part number; description; quantity; unit cost; total value.

By giving values to the items detailed in both sets of these documents, using whatever basis is decided upon – standard estimated, or actual costs – it is possible to construct a system of work in progress accounting. Normally the process layout is used as a master record and subsidiary operation cards are used to record actual direct labour times booked to the job. These can then be compared with the master record and labour efficiency calculations made and reported, so that appropriate investigative action can be set in motion.

A set of materials requisitions would be prepared from the parts lists, and if separate parts lists are prepared for each sub-assembly or component, copies of these can be adapted for use as materials requisitions. As materials are issued, these requisitions can be priced and summarised and the book entries made, debiting work in progress and crediting parts stock account. Depending on the system in force, the materials issued summary would also identify the jobs to be charged.

A system such as this provides the materials input to the work in progress account. There are, of course, ways by which materials costing systems may be made more or less sophisticated (see Chapter 7, 'Stock control and materials costing').

The work in progress labour input cost is arrived at by analysing the direct labour cost under chosen cost centres and debiting the work in progress account from the payroll summary (see Chapter 11, 'Labour costing and the use of standards'). This could easily incorporate the values assigned to the process layouts appropriate to the work carried out.

The cost of sales, or work in progress output if we assume for this purpose that there is no finished goods stock account, can often be conveniently entered on to the accounts department copies of sales invoices, the source data for this exercise being the process layouts and parts lists for the articles sold. As values will have been given to the parts and operations contained in these documents, the total materials and labour cost for each component should be readily available. By multiplying these by the quantity sold the prime materials and labour cost for each sales invoice may be determined, following which invoice batches can be summarised by product ranges, or whatever form of analysis is desired.

Example: Product *XYZ* is sold at £80 each. Costs per article include prime materials £20 and prime labour £12. During accounting month 1 materials to produce *XYZ* are issued totalling £800, direct labour bookings to this product amount to £360, and 15 are sold. In month 2 the corresponding figures are: materials issued £500, labour bookings £180, quantity sold 10. The accounting entries would appear thus:

	Debit £	Credit £
Parts stock:		
Month 1 – WIP (*XYZ*)		800
Month 2 – WIP (*XYZ*)		500
Payroll control:		
Month 1 – WIP (*XYZ*)		360
Month 2 – WIP (*XYZ*)		180
WIP materials (*XYZ*):		
Month 1 – parts stock	800	
cost of sales		300
Month 2 – parts stock	500	
cost of sales		200
WIP labour (*XYZ*):		
Month 1 – payroll control	360	
cost of sales		180
Month 2 – payroll control	180	
cost of sales		120
Cost of sales materials (*XYZ*):		
Month 1 – WIP (*XYZ*)	300	
Month 2 – WIP (*XYZ*)	200	
Cost of sales labour (*XYZ*):		
Month 1 – WIP (*XYZ*)	180	
Month 2 – WIP (*XYZ*)	120	
	2,640	2,640

The use of an analysed ledger would, of course, enable materials and labour to be separately balanced within the one work in progress account and the one cost of sales account.

MACHINE HOUR ACCOUNTING

It has been a long established practice in many branches of industry, notably printing prior to the introduction of computer technology, to cost a product or process according to the number of machine hours involved in its production.

The normal basis entails allocating overheads incurred to each of the machines employed, so that if a four-week production month was expected to incur overhead costs of, say, £13,500 and the allocation decided upon was 40 per cent to Machine 1, 35 per cent to Machine 2 and 25 per cent to Machine 3, this data would be applied to the four weeks' machine hours. Assuming an average weekly utilisation of 60 hours per machine, the calculation would be as follows:

	Overheads	M/C Hours	M/C Rate
Machine 1	5,400	240	22.50
Machine 2	4,725	240	19.69
Machine 3	3,375	240	14.06
Total	£13,500	720	£18.75

The overhead allocation would be, to a greater or lesser extent, arbitrary, although some expense items, e.g. depreciation and repairs, could be specifically apportioned, and if records were sufficiently detailed, the actual costs and machine running hours would be compared with the estimates. In some cases, the machine operators' wages would be treated as an overhead expense and included in the calculation, and in others treated separately as direct labour.

With the spread of automation, it has long been commonplace for one operator to run two or more machines simultaneously and methods have had to be devised to cope with the situation. The following example outlines one approach:

Assume that the data shown in the earlier calculation represents the period budget, and that it was planned that there would be 3 machine operators working an average of 40 hours per week at an average pay rate of £3.20 per hour.

The actual performance for the period gave these results:

(a) Machine 1 : 260 hours : overheads allocated 5,800
    Machine 2 : 180 hours : overheads allocated 4,500
    Machine 3 : 220 hours : overheads allocated 3,600

    660   £13,900

(b) Direct labour hours 500 : cost £1,700

83

This could be represented in the accounts in the following manner:

	Debit £	Credit £
Work in progress:		
(i) direct labour	1,408	
$(\frac{480}{720} \times 660 \times £3.20)$		
(ii) overheads	12,375	
$(660 \times £18.75)$		
Overhead expenditure variance:	400	
$(£13,900 - £13,500)$		
Overhead volume variance:	1,125	
$(£13,500 - £12,375)$		
Labour rate variance:	100	
$(£1,700 - £1,600,$ i.e. $500 \times £3.20)$		
Labour efficiency variance	192	
$(£1,600 - £1,408)$		
Payroll control:		1,700
(actual cost)		
Overheads control:		13,900
(actual cost)		
	15,600	15,600

So much for the accounting concepts, with this particular example illustrating unfavourable variances totalling £1,817. More is needed from the point of view of management control however, and it would probably be arranged that a reporting procedure would be built upon a system of machine time records to highlight deviations from planned utilisation.

The costing to individual jobs would involve charging a direct labour rate of $\frac{480}{720}$ or $\frac{40}{60} \times £3.20 = £2.14$ per machine hour, with overheads being charged to the job at the planned, or standard, machine hour rate.

SCRAP AND REWORK

In any system of work in progress accounting consideration must be given to the treatment of scrap and rework, or rectification. Scrap may be of two kinds. Unavoidable scrap is wastage in the form of metal cuttings, known as swarf, arising from turned parts machined from steel billets. There is also scrapped work, i.e. work scrapped in the course of production. Rework, or rectification, occurs when the product fails to pass the inspection stage at whatever operation may be involved in the manufacturing process. The goods are not scrapped in this case, but may be sent back to a previous operation to be reworked.

All these cases involve a cost of some kind and the company must ensure that sufficient allowance for that cost is made in compiling its sales prices. This responsibility is normally handled by the estimator, who will check from time to time the allowances he has built in against the accounts records of actual costs incurred.

Regarding the accounting treatment, both the input (purchase) and output (cost of sales) costs of unavoidable scrap would be accounted for as a direct charge and allocated to the product concerned.

Scrapped work involves dealing with the cost, both materials and labour, invested in work in progress up to the point of scrapping the article. In some organisations this merely takes the form of using the estimator's allowances and adding these to the unit cost of sales figures when finished products are sold.

In others a scrap reporting system is used. A sample scrap note is shown in Appendix 1 and its use, as far as accounts are concerned, is to compile a summary of the cost of scrapped work which provides the basis for adjusting the book work in progress figures, with this account being credited and an appropriate variance account being debited. The virtue of this system is that, assuming all scrapped work is properly recorded, not only can the estimator's allowances be checked, but a more accurate work in progress figure can be arrived at. The fact must be faced, however, that circumstances must dictate the methods used, and the accountant must be governed by the type of production and resources available. He would not endear himself to his production colleagues if a costly system of scrap recording and reporting revealed that scrapped work costs were minimal. Cost-effectiveness must be the name of the game.

Rework, by definition, should consist of labour costs only. Conventionally, this is dealt with by having the operators book to special codes allotted to rework, so that the times may be costed and written off as an excess cost, or variance, rather than being charged to work in progress.

14

Provisions and adjustments

PREPAYMENTS AND ACCRUALS

There are certain types of expenditure usually billed at intervals greater than the one month normally chosen as the period for which management accounts are prepared, e.g. rents, rates, insurances, telephone and bank interest.

Some of these, such as insurances, would be charged in advance, even though, in the case of this particular item, there could be premium adjustments at a later date for those risks involving declarations by the company when firm figures were available. Others, such as bank interest, would be charged in arrear, whilst quarterly telephone charges contain a mixture of both, with rental being billed in advance, and calls in arrear.

From the point of view of the users of management accounts, the charge for the month in question is what is being looked for, and just how the accountant arrives at the appropriate figures is an administrative detail with which, in the main, the rest of the management team is not vitally interested. A useful method for dealing with this type of charge is to allocate the incoming invoice to a prepaid or accrued charges control account, as appropriate, and to apportion the estimated monthly cost to the relevant expense account by journal entry. A system of standard journal vouchers, as described later in this chapter, could possibly be of use here.

Some form of analysis, to monitor the incoming charges so that it could be seen when an adjustment was required to the monthly apportionment, could be constructed on the following lines:

Accrued charges – telephone (account no.   )

Month	Detail	Period	Cost Factory 1 £	Factory 2 £	Apportionment (A/C no.) £	Balance £
Jan.	Calls	Oct.–Dec.	750			
	Rent	Jan.–Mar.	150			
			900		(500)	400
Feb.	Calls	Oct.–Dec.		600		
	Rent	Jan.–Mar.		100		
			900	700	(500)	
					(500)	
					(1,000)	600

The system caters for charges billed at intervals greater than one month. Other charges outstanding at month-end could probably just as easily be dealt with by raising a journal entry to debit the revenue account concerned, crediting the balance sheet account. These journal entries would usually be reversed in the following month when the invoices covering the outstanding charges have been recorded.

Examples might be, say, parts bought for stock, welfare supplies and vehicle servicing. In the first two instances, there could be goods inwards documentation for which suppliers invoices have not been received, and by reference to the purchase order, the goods inwards notes could be priced and totalled, while an estimate would most likely have to be made of the vehicle service cost if the garage account was not to hand.

An imaginary journal entry for items such as those outstanding at the end of Month 1 would look like this:

	Debit £	Credit £
Parts stock	1,200	
Staff welfare	80	
Vehicle expenses	150	
Accrued charges		1,430

The journal entry for the reversal appearing in the following month would, of course, appear thus:

	Debit £	Credit £
Accrued charges	1,430	
Parts stock		1,200
Staff welfare		80
Vehicle expenses		150

It is possible that there could be outstanding income after the books had closed too. This could be for retrospective selling price increases, which might even stretch back into a previous financial year. Any prior year income or expense should be allocated to a separate account for prior year adjustments, if significant.

The same principles would apply to outstanding income as for accrual charges, the difference being that, on setting up the provision, instead of crediting accrual charges, a debit entry would be made to a debtors suspense, or sundry debtors account, crediting the sales account concerned.

DEFERRED INCOME

There are occasions when the accounting treatment of deferred income has to be considered. This can arise if a sales invoice is raised by arrangement for a future charge.

The company may, for instance, provide its customers with a two years' 'free' materials and labour warranty period on certain of its products. Although not normally shown separately in the sales invoice, there will, or should, be an element of the selling price which will have been estimated to represent the warranty mark-up, and this warranty income will require to be treated differently from the rest of the sales price, as it really represents income not fully earned until the two years have expired.

A customer may perhaps receive an invoice for 5 units at £2,350 for each piece of equipment, total £11,750 plus VAT, and in arriving at this price the company could have been involved in the following sums:

Works cost	2,000
add warranty 5 per cent	100
	2,100
add profit mark-up 11 per cent	231
	2,331
Round up to	£2,350

The calculated profit per unit now becomes £250, equal to approximately 10.7 per cent of the final selling price, but this can be subject to further adjustment, whilst the warranty element of the sale of 5 units equals £500 and should be spread over a 24 months' period.

If we assume that the company operates on the basis of two periods of four weeks and one of five weeks every quarter, the income allocation of this transaction would appear thus:

	£
Month 1 (4 weeks) 5 units at £2,250	11,250
Warranty 4/104 × £500	19
	11,269
Month 2 (4 weeks) Warranty	19
Month 3 (5 weeks) Warranty £5/104 × £500	24

In this example the profit has been fully allocated to the sale of the units and no profit mark-up has been applied to the warranty income.

The revised profit percentage, excluding warranty, would therefore be $\frac{250}{2,250}$ = 11.1 per cent approximately.

In the event of any of the units requiring a warranty repair, the costs incurred would be charged in the accounts in the month in which they occur. One way of handling this would be for a 'no-charge' sales invoice to be issued, the costing copy of which, bearing the appropriate job number, would be used for the materials and labour cost of sales entry.

The situation outlined implies the keeping of some sort of running schedule, from which the warranty income apportionment would be derived. This is a comparatively simple matter, however, and could take the following form:

Year/Month billed	Total	Apportionment			
		Y1/M1	Y1/M2	Y1/M3	Y1/M4
	£	£	£	£	£
Y1/M1	1,500	57	57	73	57
Y1/M2	1,600		61	78	61
	3,100	57	118	151	118
Y1/M3	1,400			67	54
	4,500	57	118	218	172
Y1/M4	1,750				68
	6,250	57	118	218	240

Of course, the right-hand column of the schedule would be used to carry forward the unexpired balances on to a new schedule and in Year 3, Month 1, the Year 1, Month 1 figures would be deducted from the totals brought down.

Whether or not a system of income deferral is employed will depend largely on the materiality of the sums involved, and in some cases it will be appreciated that they could be quite considerable.

The same considerations apply, of course, in the case of the sale of maintenance contracts to customers.

STOCK, BAD DEBTS AND DEPRECIATION PROVISIONS

Depreciation certainly needs to be charged in the management accounts and it is often considered prudent to provide for possible losses in respect of stock and bad debts as well.

The control of stock, including work in progress, is a complex question and errors of various kinds will invariably arise. The poorer the control disciplines, the more likely it is that both the number and the seriousness of errors will increase, and the nervous accountant could be forgiven for thinking that the scope is virtually limitless. For example:

1 Errors in stocktaking, could be due to:
   (a) items being missed or duplicated;
   (b) items being wrongly described, wrong part number etc.;
   (c) work in progress being assessed at an incorrect stage of completion;
   (d) company products held by third parties not reported, and therefore missed from stock-take.
2 Errors in stock valuation due to:
   (a) incorrect pricing;
   (b) calculation errors;
   (c) customer free-issue materials wrongly included and priced.
3 'Cut-off' errors:
   (a) goods included in stock-take, but not booked in or recorded in accounts until after stocktaking date;
   (b) sales advice notes and invoices raised prior to stock-taking date for goods included in stocktaking and despatched subsequently;

(c) credit notes not issued for goods returned by customers prior to stocktaking.
4 Book-keeping errors:
    (a) suppliers' invoices wrongly allocated, e.g. repair charges allocated to stock account;
    (b) invoicing errors, e.g. incorrect totals;
    (c) goods issued from stores without materials requisitions being raised;
    (d) inadequate costing, leading to errors in cost of sales and work in progress calculations.
5 Changes in policy:
    (a) provisions for obsolete stocks etc. may be inadequate;
    (b) basis of valuation may be changed due to, for instance, changes in overhead rates.

Although not strictly errors, these can have significant bearing on discrepancies between 'actual' and book stocks.

All this underlines the case for a general stock provision adjusted, as necessary, when the results of stocktaking are known.

The accountant, in conjunction with management, would be wise to provide for possible bad debts in the management accounts. If, for example, book debts were anticipated to average, say, £600,000, it might be decided that, after taking both past experience and a forward look at the industry's prospects into account, a general provision of 5 per cent should be set up and if, at the start of the financial year, the general provision stood at £20,000, then a further £10,000 would be charged over the year in the management accounts, thus bringing the year-end total to £30,000, or 5 per cent of £600,000.

In the case of an exceptionally large bad debt occurring during the year, the situation may need to be reviewed and an additional charge made in the accounts.

Depreciation, whilst properly falling under the heading of provisions, has been dealt with in Chapter 15, to which the reader is referred for methods of depreciation, calculation bases etc.

STANDARD JOURNAL VOUCHERS

Many accountants consider it helpful to employ a system of standard journal vouchers as a means of posting the nominal ledger, in preparation for the extraction of a trial balance, from which the management accounts are compiled.

A typical journal voucher form is shown in Appendix 1, and the standard journal voucher will normally have the account names and code numbers already entered in readiness for the amounts to be filled in each month.

The journal voucher numbers designated could be on these lines, with these same designatory numbers being quoted each month:

JV1   Sales ledger transfers
JV2   Purchase ledger transfers
JV5   Payroll summary
JV10 Stores issues
JV12 Accrued purchases
JV15 Standard prepayments
JV16 Standard accounts
JV20 Stock provisions
JV25 Bad debts
JV30 Depreciation
JV40 Cost of sales
JV45 Intercompany charges
JV50 onwards. Sundry transfers

Much will depend on the type of organisation, and of course the list could include, for example, separate payroll summary journal vouchers for hourly, weekly and monthly paid employees, or even for different locations.

15

Fixed assets

VERIFICATION, CLASSIFICATION AND RECORDING

Although it is commonplace for stock in trade to be physically checked periodically, fixed assets are all too often taken for granted. Consequently, unless sound control procedures are enforced by management, the fixed assets records can be rendered inaccurate when items of plant are scrapped or transferred to another location.

Capital equipment in the books must, therefore, be readily verifiable as to existence and ownership, both internally and by the auditors.

Sensible classification and recording of fixed assets, coupled with a system of proper notification of their movement and adequate physical control procedures, will help ensure the accuracy of the records and avert the possibility of plant disposals being arranged without, for instance, knowledge of the book values.

It is often convenient, though by no means essential, to take cognisance of tax legislation when considering the classification of fixed assets. This was sometimes carried to the extent of classifying assets according to the tax allowance for each class, although this approach became somewhat outdated with the changes brought about by the 1971 and 1972 Finance Acts, with 100 per cent writing-off allowances in the year of purchase being permitted by the latter, provided that there were sufficient taxable profits.

Nevertheless, there are certain distinctions made for tax purposes, and although the paramount consideration will probably be the likely useful life of the asset category in question, the convenience in recognising the tax distinctions arises from the fact that the tax computations become easier to make, whether by the accountant or the auditors.

The relevant capital allowances (1981/82) are:

Plant and machinery, including furniture, fittings, office equipment, commercial motor vehicles and certain building costs such as expenditure to comply with fire regulations:
first year allowance of up to 100 per cent of cost.

Motor cars for business use:
writing-down allowance of 25 per cent per year of the reducing balance, with a restriction for higher priced cars.

Industrial buildings, including additions and improvements:
initial allowance of 75 per cent and a writing-down allowance of 4 per cent (of the original cost) per year.

A company might classify its fixed assets in the following way:

Category	Write-off term
1 Freehold premises	50 years
2 Freehold fixtures	50 years
3 Leasehold premises	Lease period
4 Leasehold fixtures	Lease period
5 General factory plant and machinery	10 years
6 Special purpose factory plant and machinery	5 years
7 Furniture, fittings and office equipment	10 years
8 Commercial vehicles	3 years
9 Motor cars	4 years

Although the write-off term for fixtures is the same as that shown for the same category of premises, the tax allowance would normally be different and this is the reason for keeping separate accounts. The erection of a fire escape, qualifying for a 100 per cent first year tax allowance, might be allocated to the appropriate fixtures account to simplify the preparation of accounts for tax.

If a computer facility is available, there are many software packages catering for fixed assets accounting and it is unlikely that a programme will have to be specially written. In most cases individual depreciation computations for each item of plant will be incorporated, offering the facility of instantly ascertaining written down values.

Often, however, the maintenance of a manual system will not be unduly burdensome and implementation can usually be organised with comparative ease. One approach would be to use specially ruled cards, one for each item, and these could be numbered in a special way so that all assets in a given category are in sequence. For example, if a four digit numbering system were in use, the 5000 series might indicate general factory plant and machinery and the 6000 series, special purpose factory plant and machinery.

The purchase of, say, a portable standby generator would involve the allocation of the next available number from the 5000 series, which might be, say, 5221, and this would be the number given to the card on to which the relevant information would be recorded. This same number could be painted or engraved on to the equipment itself and thus become the plant number by which the identity of the generator would be retained.

The information to be entered on each card would comprise:

(a) full description of asset, including model and serial numbers if appropriate;
(b) location;
(c) date purchased;
(d) cost;
(e) purchase invoice number or other reference;
(f) date physically checked and checker's initials;
(g) date sold or scrapped;
(h) proceeds on disposal, and
(i) cash book or other reference.

It would not normally be an essential requirement for individual annual depreciation calculations to be made, as this would be done for each asset group in total. The calculation of the written down value on disposal of an asset should present no difficulty, provided that the information entered on the card has been correctly recorded.

ACQUISITIONS AND DISPOSALS

Companies frequently find it advisable to introduce a special form to cover the authorisation of capital expenditure. Such forms can be given various titles but are often simply called capital authorisation forms, and an example of one is shown in Appendix 1.

The necessity for the control of capital expenditure by the use of a special form is due primarily to the large sums of money which may be involved and, viewed in its proper light as an investment, it must be ensured as far as possible that there will be an adequate return. It is, nonetheless, surprising how frequently cases can be found of lack of proper control in this area in large as well as small organisations, with the result that the chances of successful investments are lessened and the risks of cash flow problems increased.

Capital authorisations should apply to all expenditure of a capital nature and would exclude equipment being leased. The CA form should contain the following details:

(a) date of proposed expenditure;
(b) description of equipment to be purchased, and asset category;
(c) whether equipment is ancillary to existing plant;
(d) department and area in which to be installed;
(e) supplier;
(f) external cost of acquisition, including carriage and installation and whether alternative quotation available;
(g) internal costs to be incurred;
(h) estimated useful life of asset;
(i) investment justification;
(j) whether proposed expenditure renders existing equipment redundant, and disposal plans;
(k) warranty/maintenance contract details;
(l) details of external authority approval required, if applicable, and
(m) budget reference.

It may be that the installation of a certain piece of plant will require weekend working by the maintenance department and these overtime costs, together with any other internal costs involved, such as materials drawn from stores, would be itemised and a summary given under heading (f). These costs would be additional to these invoiced by the supplier under (e) and form part of the total acquisition expenditure and even though company policy may dictate that in-house labour costs are not to be capitalised, the total outlay obviously needs to be known. Without reference to a check list, it is only too easy to omit consideration of likely internal costs from the equation. Investment justification is reviewed in 'Buy or lease decisions' later in this chapter.

Where applicable, details of any warranty period, together with information regarding appropriate maintenance contract choices must be checked and the results noted on the CA form. These data are relevant, both to assure that once acquired, the plant will remain operational, and to bring this element into the calculation of operating costs.

External authority, e.g. local council, approval may be needed in special cases where, for example, the plant may be required to meet pollution control standards, and in other cases it may be necessary to obtain a licence to operate the plant, or the process to be operated by the plant. The CA form should signify whether appropriate approval has been obtained or has been applied for, where relevant. It should also indicate whether the proposed acquisition was included in the capital budget and, if so, the budget item number.

A possible CA routing arrangement could be on the following lines:

Originator (keep copy)
Department head
Accountant (keep copy) – to complete item (i), investment justification
Managing director

Following approval, which may be preceded by a board meeting, duly signed copies of the form to be distributed to the departmental head, the accountant and the company secretary (for insurance purposes).

It may be arranged that capital items below a certain cost level be grouped, thus avoiding the need for individual CA forms. It could, for instance, be decided that items costing less than £100 would not be capitalised, but written off as purchased. Items over £100 in value, but costing less than £1,000, could then be grouped, with the managing director allocating a sum covering a stated period of time for each department to spend on these. Individual acquisitions of £1,000 or more would require separate CA forms.

A capital authorisation control file would be kept by the accountant, from which updated summaries could be prepared for submission to management. These summaries would give authorised expenditure by department and category, as well as costs actually incurred and outstanding commitments. Budget variances should also be shown, so that it could easily be seen which departments were overspending and those which had not committed their budget appropriations.

Even though it is not usual for a special form to be used when fixed assets are scrapped or sold, the importance of exercising control over such events must not be overlooked. Many instances have been known of managers scrapping plant without arranging for the accountant to be informed, usually because the manager is aware of the age of the plant and the fact that it has been written off.

It must be stressed to those entrusted with the disposal of fixed assets that it is essential that all disposals must be properly notified, so that the records can be corrected and the relevant asset account credited, with the depreciation provision account being debited and any difference between the two allocated to a profit or loss on disposal of fixed assets account. Any proceeds on disposal would, of course, be credited to this latter account.

It is common practice for instructions to be given that disposals of fixed assets can only be made with the express approval of the managing director, with approval being given or withheld after obtaining relevant data from the accountant, e.g. net book value, whether there is a hire purchase liability etc.

For the system to operate effectively, there must be no confusion over exactly what is being disposed of and the notification must, therefore, be quite explicit in order that the item concerned can be correctly identified in the fixed assets register.

DEPRECIATION

Opinions vary as to how best to provide for the depreciation of fixed assets. A number of methods can be found in accountancy text books, but the straight line and reducing balance methods are popularly used in practice, with the former being more generally preferred.

This entails deciding upon the number of years over which assets of a particular category are to be written off in the accounts, calculated from date of purchase, and charging equal annual instalments against profits in order that, at the end of the predetermined period of time, the depreciation provision includes an amount equal to the original cost of the asset, i.e. the net book value is nil.

The reducing balance method usually involves higher percentage rates of depreciation with the important difference that the calculation is based on the written down value of the asset, as opposed to its cost. The depreciation charge is, therefore, heavier in the earlier years, tailing off as the age of the asset increases and equalisation is theoretically achieved by a converse pattern of repairs, the cost of which would normally be expected to rise as the asset ages. This method does, however, possess a distinct disadvantage in so far as the asset is never fully written off and this often has to be overcome by fixing bottom limits at which residual balances are dealt with by being added to the depreciation charge normally calculated.

Although inflation accounting is referred to in Chapter 17, it is worth mentioning at this point that a growing number of companies have turned to replacement cost accounting. This approach recognises that the replacement of an asset will invariably cost considerably more than the original and it is asserted that additional depreciation should be provided, so that when added to the depreciation provision computed under the historical cost method, an amount nearer to the replacement cost will have been set aside out of profits, thereby in an extreme case possibly even averting the need to raise additional capital.

The additional depreciation thus provided would usually be credited to a replacement cost reserve account and this practice, or others similar, has, or have, been followed by many businesses for a number of years, even before the introduction of the current cost accounting standard, SSAP 16, in 1980.

When such systems are used, however, accountants and businessmen must be alive to the implications of technological change and be prepared to exercise a degree of discernment. The original equipment, for example, may not be replaced by an identical model, or may not even be replaced at all, and changed market conditions also may have an impact on replacement decisions.

Furthermore, there are well known instances of technically superior machines appearing on the market at a fraction of the cost of machines formerly carrying out similar functions. Witness the dramatic transformation in the field of electronic calculators in the 1970s. In the late 1960s a non-printing desk top model cost nearly £500 and pocket models were unknown. Less than ten years later, pocket models were available starting at prices under £10 and desk model printing calculators for less than £100.

There are, then, many influences to be taken into account and it is unwise to adopt a blind 'across the board' approach to replacement cost accounting.

Whichever depreciation method is used, the accountant must decide its treatment in the accounts and will usually find management receptive to the advice that his expertise enables him to offer, in order to assist in the formulation of sensible policies.

Commonly, a full year's depreciation is charged in the year of purchase and none in the year of sale, but the application of this procedure can create problems. Imagine a machine costing £100,000 being purchased in the last month of the financial year, with depreciation at 10 per cent per annum straight line. If the practice in calculating depreciation for the management accounts is to base this entirely on the cost of the assets in the books, adjusted for movements, an additional charge of £10,000 would suddenly pop up in the last month.

An acceptable alternative is to charge the budgeted figure in the management accounts, reviewing the position quarterly, or half-yearly, in conjunction with management and taking account of existing and probable major capital expenditure budget variances. Following such a review an adjustment would be made to the depreciation charge, if thought necessary.

A further variation would be to charge a full year's depreciation in the year of purchase, only if purchased during the first half-year and none if purchased in the second half, with depreciation being charged in the year of sale if sold in the second half, no depreciation charge being made on sales in the first half-year.

This can have a significant effect on the written down value of an asset. As an example, a comparison can be made in the case of a vehicle costing £8,000 purchased in December, Year 1, being sold in January, Year 4, with the financial year ending 31 December.

	Method 1	Method 2
	£	£
Cost	8,000	8,000
Depreciation – Year 1	2,000	–
Year 2	2,000	2,000
Year 3	2,000	2,000
Year 4	–	–
	6,000	4,000
Written down value	2,000	4,000

In both cases of course, the vehicle has been used for just over two years, but one method shows a net book value double that of the other.

Depreciation policy is not, therefore, simply an academic question and book values have to be taken into account when considering asset disposals. Nevertheless, common sense dictates that management should not be inhibited from replacing fixed assets if the circumstances demand this to be the proper course of action.

BUY OR LEASE DECISIONS

The term 'project appraisal' is usually applied to the process of weighing the pros and cons of plant investment decisions. There are many factors to be taken into account, and accountants and businessmen will invariably be involved in a degree of judgement making.

It is an age-old maxim that the higher the degree of risk, the higher the expected profit and judging these factors obviously implies an attempt to quantify them in cash terms. Likely technological and market changes have to be deliberated and investment in North Sea oil exploration, for example, would normally be thought to be a riskier venture than, say, investment in food processing plant. The latter would, therefore be expected to show a correspondingly lower return.

In certain situations there may be a government grant available, such as under the Electronic Components Industry Support Scheme, or in prescribed regional development areas, which should be included, where relevant.

Managers and accountants must also see that each is aware of the relevant technical data implications of plant productivity and operating costs. Not least of the complex questions to be considered in addition to likely market conditions regarding the product are the economic outlook generally, and the tax position. Further mention will be made of these following the examples given.

There are various techniques available by which alternative courses of action may be measured. Chief of these are (a) the pay back method; (b) the average return on capital method, and (c) the discounted cash flow method.

Pay back

This technique takes into account the estimated profits arising from the projected investment and is used to calculate the period of time required to elapse before the profits amount to a sum equal to the cost of the investment. It does not take account of the cost of money, i.e. interest, or of profit earned after the pay back period.

Average return on capital

This is simply a calculation of the total expected profits over the life of the investment, expressed as an average percentage annual return. It also does not take the cost of money into account and ignores the timing factor, so that it is possible for a short-life asset to give the same return as one with a longer life still earning profits.

Discounted cash flow

This technique recognises the cost of money and thereby brings interest into the calculations by converting future expenditure and earnings into present values.

Buy or lease decisions can be necessary in a variety of situations, whether it is a computer or photocopying facility contemplated, a vehicle fleet, premises, or manufacturing plant. The same sort of process has to be gone through and the sums worked, even though with so many factors to be permutated, it is highly likely that subsequent events will disprove at least some of the estimates made.

Nevertheless, remembering that all decisions contain an element of judgement, it must be better to have as much relevant information as possible so that, hopefully, the judgements may be sounder.

In the examples on pp. 104–7, the following data are assumed:

1 An existing machine produces an average of 5 units per week for 50 weeks of the year, giving a total annual quantity of 250. These sell at £250 each and realise a profit of 10 per cent, i.e. £6,250 per annum.

2 A newer machine costing £75,000 is available, capable of double the output rate and its acquisition is contemplated. The labour operating costs are identical and it is thought that continuing market demand is such that the unit selling price of £250 can be maintained, but with a proportionately lower fixed overhead content, the profit percentage is increased to 15 per cent, equivalent to £18,750 per annum.

3 The newer machine is thought to have a useful life of 8 years minimum, with a residual value of £2,500, and by an annual increase in maintenance expenditure of £250, starting at £1,500 for Year 1, it is thought that the existing machine could be kept operational for this length of time.

4   At Year 1 prices, it is reckoned that the newer machine would entail an average annual expenditure of £200 in respect of tool repairs and consumable supplies, as against £300 already being incurred with the existing machine.
5   There is also the possibility of acquiring the newer machine by means of a hire purchase agreement over 3 years, with a flat rate of interest at 17 per cent per annum.
6   If it is decided to buy the newer machine, whether or not by hire purchase, there would be a free maintenance period of 2 years, followed by a maintenance contract for 3 years at a cost of £1,000 per annum renewable for a further 3 years at £2,500 per annum.
7   There is a further possibility of leasing the newer machine involving a fully inclusive maintenance hire charge of £20,000 per annum for 5 years, renewable at £5,000 per annum.
8   An inflation allowance, after Year 1, of 12 per cent per annum is to be incorporated in the computations, the cumulative effect being to multiply Year 1 prices by the following factors:

Year 2	1.120
Year 3	1.254
Year 4	1.405
Year 5	1.574
Year 6	1.762
Year 7	1.974
Year 8	2.211

9   Average annual interest is reckoned to be 15 per cent, giving present values from interest tables for each £ as follows:

Year 1	0.870
Year 2	0.756
Year 3	0.658
Year 4	0.572
Year 5	0.497
Year 6	0.432
Year 7	0.376
Year 8	0.327

Example A

Retain	Year 1	Year 2	Year 3	Year 4	Year 5	Year 6	Year 7	Year 8
	£	£	£	£	£	£	£	£
Maintenance	1,500	1,750	2,000	2,250	2,250	2,750	3,000	3,250
Tool repairs and consumable supplies	300	300	300	300	300	300	300	300
Profit generated	(6,250)	(6,250)	(6,250)	(4,688)	(4,688)	(4,688)	(3,125)	(3,125)
	(4,450)	(4,200)	(3,950)	(2,138)	(1,888)	(1,638)	175	425
Inflation 12 per cent	–	(504)	(1,005)	(866)	(1,083)	(1,249)	170	515
	(4,450)	(4,704)	(4,955)	(3,004)	(2,971)	(2,887)	345	940
Cumulative	(4,450)	(9,154)	(14,109)	(17,113)	(20,084)	(22,971)	(22,626)	(21,686)
Present values (15 per cent discount)	(3,872)	(3,556)	(3,260)	(1,718)	(1,477)	(1,247)	130	307
Cumulative	(3,872)	(7,428)	(10,688)	(12,406)	(13,883)	(15,130)	(15,000)	(14,693)

Example B

Buy	Year 1	Year 2	Year 3	Year 4	Year 5	Year 6	Year 7	Year 8
	£	£	£	£	£	£	£	£
Plant purchase	75,000							
Maintenance contract		–	1,000	1,000	1,000	2,500	2,500	2,500
Tool repairs and consumable supplies*	200	200	200	200	200	200	200	200
Profit generated*	(18,750)	(18,750)	(18,750)	(18,750)	(18,750)	(18,750)	(18,750)	(18,750)
Residual value								(2,500)
	56,450	(18,550)	(17,550)	(17,550)	(17,550)	(16,050)	(16,050)	(18,550)
*Inflation 12 per cent	–	(2,226)	(4,712)	(7,513)	(10,648)	(14,135)	(18,068)	(22,464)
	56,450	(20,776)	(22,262)	(25,063)	(28,198)	(30,185)	(34,118)	(41,014)
Cumulative	56,450	35,674	13,412	11,651	39,849	(70,034)	(104,152)	(145,166)
Present values	58,861(A)	(15,707)	(14,648)	(14,336)	(14,014)	(13,040)	(12,828)	(13,412)
(15 per cent discount)								
Cumulative	58,861	43,154	28,506	14,170	156	(12,884)	(25,712)	(39,124)

(A) Assumed plant purchased Day 1 and therefore not discounted.

Present value thus   75,000

less 0.870 × 18,550   16,139

                      58,861

Example C

Hire purchase	Year 1	Year 2	Year 3	Year 4	Year 5	Year 6	Year 7	Year 8
	£	£	£	£	£	£	£	£
Plant purchase:								
Principle 75,000								
Interest (17 per cent p.a. flat) 38,250								
113,250	37,550	37,550	37,550					
Maintenance contract	–	–	1,000	1,000	1,000	2,500	2,500	2,500
Tool repairs and consumable supplies*	200	200	200	200	200	200	200	200
Profit generated*	(18,750)	(18,750)	(18,750)	(–8,750)	(18,750)	(18,750)	(18,750)	(18,750)
Residual value								(2,500)
	19,200	19,200	20,200	(17,550)	(17,550)	(16,050)	(16,050)	(18,550)
*Inflation 12 per cent	–	(2,226)	(4,712)	(7,513)	(10,648)	(14,135)	(18,068)	(22,464)
	19,200	16,974	15,488	(25,063)	(28,198)	(30,185)	(34,118)	(41,014)
Cumulative	19,200	36,174	51,662	26,599	(1,599)	(31,784)	(65,902)	(106,916)
Present values (15 per cent discount)	16,704	12,832	10,191	(14,336)	(14,014)	(13,040)	(12,828)	(13,412)
Cumulative	16,704	29,536	39,727	25,391	11,377	(1,663)	(14,491)	(27,903)

Example D

Lease	Year 1	Year 2	Year 3	Year 4	Year 5	Year 6	Year 7	Year 8
	£	£	£	£	£	£	£	£
Hire charge	20,000	20,000	20,000	20,000	20,000	5,000	5,000	5,000
Tool repairs and consumables*	200	200	200	200	200	200	200	200
Profit generated*	(18,750)	(18,750)	(18,750)	(18,750)	(18,750)	(18,750)	(18,750)	(18,750)
	1,450	1,450	1,450	1,450	1,450	(13,550)	(13,550)	(13,550)
*Inflation 12 per cent	–	(2,226)	(4,712)	(7,513)	(10,648)	(14,135)	(18,068)	(22,464)
	1,450	(776)	(3,262)	(6,063)	(9,198)	(27,685)	(31,618)	(36,014)
Cumulative	1,450	674	(2,588)	(8,651)	(17,849)	(45,534)	(77,152)	(113,166)
Present values (15 per cent discount)	1,262	(587)	(2,146)	(3,468)	(4,571)	(11,960)	(11,888)	(11,777)
Cumulative	1,262	675	(1,471)	(4,939)	(9,510)	(21,470)	(33,358)	(45,135)

Summary
Cumulative net (inflows) outflows

	Example A Retain	Example B Buy	Example C Hire-purchase	Example D Lease
	£	£	£	£
Year 1	(3,872)	58,861	16,704	1,262
Year 2	(7,428)	43,154	29,536	675
Year 3	(10,688)	28,506	39,727	(1,471)
Year 4	(12,406)	14,170	25,391	(4,939)
Year 5	(13,883)	156	11,377	(9,510)
Year 6	(15,130)	(12,884)	(1,663)	(21,470)
Year 7	(15,000)	(35,712)	(14,491)	(33,358)
Year 8	(14,693)	(39,124)	(27,903)	(45,135)

Conclusions

In this particular case, leasing proves to be the most favourable course over the full period of 8 years, producing a total net inflow of £45,135, with buying next at £39,124 and Examples A and C well down the line.

Examination of the yearly cash flows shows that, in the early years, buying would prove to be a considerable drain on resources and if bank borrowing needed to be kept in check, for instance, leasing might be even more strongly favoured.

However, Examples B and C will be affected by tax, with the full capital allowance against the investment of £75,000 being available in Year 1, subject to sufficient taxable profits and bringing this factor into account could sway the decision in favour of buying.

Further, in these examples it has been assumed that productivity and profits are to remain constant at Year 1 prices and if this pattern were to be changed, it will be appreciated that this could significantly affect the outcome.

If it had been thought that it was not possible to retain the existing machine due, for example, to the non-availability of replacement parts, then of course, this option would not be available. On the other hand, if it was thought that the machine could be kept going for another 3 or 4 years, by which time a superior model would be on the market, then retention would seem to offer the best return. Market conditions, however, could dictate otherwise and in order to satisfy demand it might still be decided to lease.

If the existing machine could not be retained, leasing would only prove to be more favourable over 4 years, if termination did not involve too heavy a penalty.

General economic conditions will influence the interest and inflation rates chosen, and in the examples given both have been assumed to be constant over the full 8-year period.

If plant investments are normally made with the aid of bank borrowings, and hire purchase is considered as an alternative, it could be argued that, with the investment being made at the start of Year 1, a fairer comparison would be made by increasing the plant purchase figure of £75,000 in Example B by an amount representing overdraft interest.

Finally, it will be realised that the examples given do not advocate a particular policy for every case which might arise. The circumstances will invariably be different and it is important to collect and assess the data before making the calculations, and only then can the various alternatives be objectively weighed.

16

Cost control with budgets and standards

BUDGET PREPARATION AND RESPONSIBILITIES

In most businesses the accountant is charged with the responsibility of co-ordinating the work carried out by his fellow managers in preparing preliminary budgets for their individual areas of influence, as well as collating the results for presentation to his managing director, together with such budget papers as he has separately prepared, such as finance charges and cash flow statements.

Each manager must be clearly aware of the responsibilities assigned him and it is of paramount importance, if meaningful budgets are to be constructed, that in appropriate cases his subordinates are involved in setting, and accepting responsibility for, realistic budgets for their sections in terms of materials usage and labour efficiency. This is the essence of effective budgetary control, which implies that, in addition to being a planning document, the budget is also used as a forward looking management control tool, in sharp contrast, for example, to the comparison of current with past performance.

This is not to say that the budget should be allowed to become an obsession and major changes in circumstances, not envisaged at the time of setting the budget, may have to be kept in mind if it is not feasible to cope with the extra work load involved in revising budgets. Retaining a proper sense of perspective will enable intelligent use to be made of the budget in comparing performance with plan.

Budget preparation often affords the accountant an ideal opportunity of learning more about the business from his management colleagues, examining the reporting format to see whether it might be advantageous to make changes, and demonstrating how more effective use of the accounting services can help the production manager in the solution of some of his problems, e.g. by highlighting variances and thereby indicating areas where action may be required.

This is known as management by exception and is an important aid to managers in fulfilling their responsibilities, enabling them to concentrate on problem areas and pay less attention to those parts of the activity which are within budget.

The compilation of departmental budgets frequently discloses areas of conflict, notably between production and sales, and the accountant should undertake initial responsibility for resolving the differences by meeting with sales and production management and pointing out the differences. It might be possible to increase sales by more advertising if there is an apparent surplus of production capacity, or in the event of a capacity shortage, perhaps more labour could be recruited. The managing director would normally take responsibility as final arbiter in the case of continued disagreement.

A useful practice is to circulate preliminary budget forms to managers, listing the appropriate income and expenditure headings and requesting the insertion of budget figures. For information, current operating figures could be entered on the forms prior to circulation, and where weekly data are issued, such as sales or payroll reports, these could perhaps be used in conjunction with subordinate managers, as a guide in forecasting monthly and annual totals. The preliminary information requested from management should include itemised projected capital expenditure, as well as manpower budgets by sections, detailing any proposed headcount changes.

It is common practice for sales to be chosen as the first stage in the budgeting process and a start is usually made by analysing the order book by month and product group. To the totals thus derived must be added reasoned estimates of achievable orders deliverable within the budget period. This can be done by reviewing what has been achieved and taking a view as to what is possible and likely, taking cognisance of the market and of products susceptible to demand changes due to external influences, e.g. anticipated higher energy costs. The table on page 112 shows a typical budget work programme.

Such a programme should then allow sufficient time for further revisions and copies to be made and distributed before the end of Month 12, in readiness for use at the start of Month 1.

Budget	Responsibility	Deadline	
Sales	Sales manager	Week 3	Month 10
Sales department overheads	Sales manager	Week 4	Month 10
Purchasing programme	Buyer	Week 4	Month 10
Direct labour	Production manager	Week 4	Month 10
Gross margins	Accountant and estimator	Week 1	Month 11
Stocks	Accountant and buyer	Week 1	Month 11
Production overheads	Production manager	Week 2	Month 11
Administration	Accountant and managing director	Week 2	Month 11
Consolidated capital budget	Accountant	Week 3	Month 11
Cash flow	Accountant	Week 3	Month 11
Draft summary (for MD's approval)	Accountant	Week 4	Month 11
Revised summary (for board approval)	Accountant	Week 1	Month 12

The construction of a simple gross margins budget is illustrated below:

	Product A	Product B	Product C	Total
Quantities	1,200	1,600	2,500	
	£	£	£	£
Unit prices				
Sales	1,000	500	200	
Materials and sub-contract	300	165	80	
Direct labour	100	60	30	
	£000	£000	£000	£000
Totals				
Sales	1,200	800	500	2,500
Materials and sub-contract	360	264	200	824
Direct labour	120	96	75	291
Gross margins	720	440	225	1,385
Percentage sales	60.0	55.0	45.0	55.4

	Product A	Product B	Product C	Total
Sales				
Month 1	200	150	60	410
Month 2	200	150	80	430
Month 3	300	250	70	620
Month 4	250	100	70	420
Month 5	100	75	100	275
Month 6	150	75	120	345
Totals 1st half-year	1,200	800	500	2,500
Gross margins				
Month 1	120	82.5	27.0	229.5
Month 2	120	82.5	36.0	238.5
Month 3	180	137.5	31.5	349.0
Month 4	150	55.0	31.5	236.5
Month 5	60	41.3	45.0	146.3
Month 6	90	41.2	54.0	185.2
Totals 1st half-year	720	440.0	225.0	1,385.0

BUDGET PERIOD AND REVISIONS

The financial year is the period normally chosen for the formulation of the operating plans to be embodied in the company budget, although it is sometimes necessary to prepare a revised budget, perhaps at the half-year stage, before the original budget has run its course.

This could be for one or more of a number of reasons not allowed for at the time of setting the original plan, e.g. trade recession, major reorganisation, strike, unforeseen boom. Only if these factors are likely to have a major impact should a revision, particularly a full-scale one, be undertaken, in view of the considerable amount of work entailed.

Although it is widely recognised that there are occasions when it makes sense to re-budget, confusion can be caused when it comes to the question of year-to-date budget figures after a revised budget has been introduced. A decision must be made as to whether the original budget is to be substituted by 'actual' figures for the period prior to the revision, so that the cumulative revised totals comprise 'actuals' plus new budget. Alternatively the original and new budgets for their respective periods can be added together to provide the revised budget for the year.

This latter method could be inappropriate in certain types of businesses. Imagine a situation in which a revised budget is to be implemented with effect from Month 7. If the original sales budget included the invoicing of a major order in Month 7, and progress was such that it was ready ahead of schedule and the customer agreed to the delivery and invoicing being made in Month 6, then this item would not appear in the cumulative budget, which could give rise to a gross distortion.

Whichever method is chosen, the decision must be clearly communicated to prevent confusion if the correct conclusions are to be elicited from a comparison of performance with plan.

Companies are increasingly becoming involved in long-range planning for periods exceeding the normal budget year, the annual budget becoming the first year of the long-range plan, otherwise described as the corporate plan, with a further year being added at the far end of the plan. Five years is a typical time scale chosen for a corporate plan.

Because such plans are long term by nature they may not be as detailed as the annual budget but, of course, the key data must be shown, and corporate planning implies a thorough review of objectives and the strategic choices available to fulfil these objectives.

ESTABLISHING AND UPDATING STANDARDS

For those businesses whose activities lend themselves to standard costing – certainly a large proportion of manufacturing industry – this method of costing forms the 'nuts and bolts' of a sound budgetary control system.

Setting standards involves close co-operation between buyers, planning engineers, and accountants and, if diligently carried out, almost inevitably leads to a greater understanding by each of the problems and requirements of the others, and thence to a better understanding of the business generally. The execution of any plan must be enhanced if there is a greater chance of the people involved consciously pulling in the same direction.

The time scale over which the standards are to be operative must be determined before standards can be set and a view taken of the average price/wage level over that period.

If, say, in setting a materials standard for aluminium for a six-months period, the price at the time of setting was £815 per tonne, and it was estimated that prices would rise by 6 per cent over the period, it could be decided to apply an average 3 per cent mark-up to allow for inflation and to fix the standard at £839.45 per tonne.

Month 1 purchases at, say, £818 per tonne would result in a favourable price variance of £21.45 per tonne, with £839.45 per tonne being allocated to stock and charged to the job, and the favourable price variance being shown separately in the management accounts as a reduction of the cost of sales total.

Should the price be £850 in Month 5, exactly the same procedure would apply, except that in this case there would be an unfavourable price variance of £10.55 per tonne and this would be added to cost of sales. Unless buying and/or inflation followed a markedly different pattern from that envisaged, the tendency would be for favourable price variances in the earlier months to be negated by unfavourable variances later, hopefully resulting in only minor differences over the period as a whole.

So much for the costing in of aluminium and we must now look at the cost of sales, or costing out process.

If 2 kilograms of aluminium were needed for the manufacture of sub-assembly XYZ, the standard cost of aluminium per sub-assembly would be £1.68 and if, in the absence of an effective scrap reporting system, it was reckoned that a scrap allowance of 5 per cent should be added, the standard would be set at £1.74.

The cost of any other materials and sub-contract charges applicable to the sub-assembly would be compiled in a similar fashion and, when added to other sub-assembly and final assembly materials and sub-contract costs, would form the total unit prime materials and sub-contract cost of the final product. Assuming this total standard, inclusive of contingencies, to be £80, this would be the sum to be debited to the materials cost of sales account and credited to work in progress materials, or finished goods account, for each final product sold.

In some businesses, notably process industries, it is less easy to calculate with precision the amount of materials used in the final product. Efficient process regulation and control obviously helps, but in the case of a plating operation, for example, although sophisticated formulae may be employed to take account of such factors as differences in component and batch sizes, plating density, agitation, temperature and evaporation, the quantity and cost of the metal finally deposited on the component frequently remains a highly educated guess at the end of the day.

The theoretical value of the plating tank stock can be calculated, using a system of full or partial standard costing, but will invariably differ from the value arrived at from the reading taken from the tank meters. Such differences would be evaluated and allocated to a materials usage account.

The procedure for establishing labour standards is basically the same as for materials. As well as a prediction of hourly departmental pay rates, standard job times are set and sometimes adjustments made for changes in production methods, 'learning curves' etc.

Differences between standard and actual departmental pay rates are costed as labour rate variances and between standard and actual job times as labour efficiency variances (see Chapter 11, 'Labour costing and the use of standards').

Although standard overhead rates can also be fixed, the method depending on the overhead recovery system used, this need not imply an overhead allocation to each product.

Truly it has been said that no product, or even department, makes a profit, only the business. This view is often reflected in the overhead allocation basis adopted, which will naturally vary according to the type of business and will often result in the application of standard overhead rates to the months' figures in total, company, departmental, or otherwise, stopping short of allocating overheads to individual products and recognising that the volume of business is the primary factor governing the proper level of overhead recovery.

There are two types of overhead variance most commonly reported: expenditure or cost, and volume or recovery, and the following is one way in which they can be presented.

Assume Department A's overhead budget for the month in question was £50,000 and that, with the overhead allocation being based on direct labour cost, the budgeted percentage was 250 per cent. Actual overheads totalled £48,000 and the direct labour input at standard was £18,000.

	£
Actual overheads	48,000
Standard overheads (18,000 × 250 per cent)	45,000
Net unfavourable variance	3,000

Analysed thus:	£
Unfavourable volume variance	
(50,000 – 45,000)	5,000
Less favourable expenditure variance	
(50,000 – 48,000)	2,000
	3,000

The frequency with which standards (materials, labour or overheads) are updated will depend upon the circumstances of the case, as will the extent. A full-scale standards revision is often time consuming and, therefore, costly and for this reason many firms are reluctant to embark on the exercise more than once a year.

Naturally, there are occasions when a re-think may be necessary, even essential, if a major change, or series of changes, occurs in the business position, e.g. the closure of a major supplier with a consequent series of 'knock-on' effects. On the other hand expedients, falling short of a full-scale revision, may sometimes satisfactorily be adopted.

It may be that only labour pay rates are to be changed, with labour times and materials standards remaining unaltered. It often happens where payment by results (PBR) schemes are in operation, however, with the standard time forming the basis for the incentive payment, that there is no alternative but to change labour standards every time there is a change in production methods.

Sometimes an attempt is made to cater for inflation by adding, at a given point during the year, an 'across the board' percentage to materials standards. The danger with this approach is that, as well as changes taking place in buying volume patterns, the incidence of inflation could be violently uneven, with some parts price increases of 5 per cent running parallel with others at 30 per cent and maybe one or two even showing price reductions. There is seldom an easy answer and the situation has to be judged in the light of sometimes conflicting factors.

VARIANCE ANALYSIS

A wealth of information can be generated when analysing variances of all types and, in attempting to be helpful, the accountant must be wary of unwittingly creating confusion among his colleagues. Variance analysis is, however, one of the most important areas of management control of the business and it is vital that timely information be issued and understood.

Responsible management will take a keen interest in the data and occasionally suggest changes in presentation. Only a constant dialogue between the accountant and the factory manager will ensure that the maximum benefits are derived from the reporting system.

Budget variances will be apparent from a properly structured set of management accounts, but 'back up' information may be called for in certain situations. Although it should be obvious from an overhead statement which expense headings are exceeding budget, and these could be significant enough to require investigation and explanation, it is not always safe to assume that all is well even with items within budget.

Pay increases may be higher than budgeted and yet total payroll costs could be within budget if the company is under staffed. This would perhaps be tolerable if a rearrangement of the work load was possible, thus enabling the business to meet its objectives. On the other hand, if the only feasible alternative was to recruit, and perhaps train, more staff, payroll costs could very soon exceed budget.

Gross margin budget variances can usually be highlighted by differentiating between volume and mix. In the gross margins budget illustrated earlier in this chapter, Month 1 showed the following data:

Product	Quantity	Sales value £	Gross margin £
A	200	200,000	120,000
B	300	150,000	82,500
C	300	60,000	27,000
		410,000	229,500

If the actual unit sales and gross margin values were as budgeted but the quantities sold were 100 of A, 400 of B and 600 of C, the results would be summarised thus:

Product	Quantity	Sales value £	Gross margin £
A	100	100,000	60,000
B	400	200,000	110,000
C	600	120,000	54,000
		420,000	224,000

An increase in sales, therefore, of £10,000 produced a gross margin of £5,500 less than budget. The budgeted margin of £229,500 represented 56.0 per cent (approximately) of sales and if the same mix had been achieved, this should have resulted in a gross margin increase of £5,600.

The gross margin variance can then be shown as:

Volume variance	5,600	favourable
Mix variance	(11,100)	unfavourable
Net variance	£(5,500)	unfavourable

An appraisal would probably be made to determine whether the sales shortfall of Product A was due to supply or demand problems. In the former, could more production capacity be made available? If the latter, what action could reasonably be taken to stimulate demand?

The situation could possibly be retrieved in the following months by maintaining and satisfying the increased demand for Product C.

The analysis of materials price variances can take the form of a print out or listing of the relevant purchase invoices. The manager concerned is then able to identify significant items from his copy of the listing and is in a position to ask for further information or be supplied with copies of invoices.

Separate listings could be prepared for each product group or by supplier, according to circumstances and, if the facility is available, an improvement could be made by the inclusion of such data as (a) part number; (b) quantity; (c) unit price; (d) standard; and (e) total variance, thus saving valuable management time in assimilating the information.

Price variance information is vital in controlling the company's operations. Can savings be made by obtaining competitive quotations? Can price increases be passed on? Are alternative materials available? These are some of the questions demanding answers when unfavourable price variances begin to assume significant proportions.

In some cases price variance data can be issued, even before the arrival of suppliers' invoices, by analysing copies of purchase orders, although it will be realised that, depending on the timing of deliveries, this will probably represent advance information and may not, therefore, agree with the monthly accounts totals.

When a standard costing system is used, materials scrap is normally reported as a variance. The materials scrap variance could be the value of the receipts from scrap sales less the materials cost of scrapped work. With usually fewer items involved than in the case of price variances, the analysis of materials scrap variances is a comparatively simple task, but should not be overlooked.

Labour rate variances for each department or section should be automatically generated by the payroll reporting system. Assuming a reasonably sensible departmental breakdown, little analysis should be required. Occasionally, however, it may happen that an abnormally large variance occurs and, unless the reasons are already obvious to all concerned, an investigation may be required.

The important point to be borne in mind is the determination of whether or not the abnormal variances are temporary in nature. Variances of a permanent nature, arising perhaps from larger than budgeted pay increases, imply that the standards are incorrect and may call for adjustments to selling prices, whereas temporary abnormalities would not usually activate selling price changes.

The analysis of labour efficiency variances is best reported in terms of job times rather than labour cost, although of course these costs will be reported in the accounts. An efficient production management team, however, will be keenly interested in job times and alert to the implications of unacceptable variances, and the reporting system, therefore, should be designed to provide production managers and estimators with early and regular data showing job and operation numbers, quantities, standard times and times taken.

In addition, details of the operators concerned may also be required and it is a good idea to provide space on the job card or time sheet for supervisors' comments, giving, for example, the reasons for excess times, such as poor tooling, slow operator etc.

Rework, or rectification, is also often reported as a labour variance and sometimes added together with the labour cost of scrapped work to form a labour scrap and rework variance.

Recording and reporting such costs in this way enables checks to be made on the estimators' allowances and future estimates can thus be adjusted, as appropriate. The analysis of such variances, would, of course, be by job or part numbers.

GENERATING ACTION

A fundamental precept in any system of budgetary control and standard costing is that there should be an awareness by all concerned of progress, measured by generally accepted yardsticks made against the agreed plan of action, and that deviations from plan must be highlighted, especially to those with particular responsibility.

For these deviations, or variances, to be investigated and acted upon clarity in reporting is obviously essential. Lack of clarity will, unfortunately, often lead to the variances being obscured and the need for action may not be immediately apparent.

An attempt has been made to illustrate some of the questions posed by variance analysis with a view to providing answers as a basis for action. The reporting system should not be too burdensome, however. It should be as simple as circumstances allow, acknowledging that managers have to read, digest and act upon the reports.

Managers, including accountants, have a duty to establish a rapport with their colleagues. It is surprising how often they simply fail to talk to one another, with the solution of problems going by default and sometimes leading to avoidable losses. For this reason, even if the reporting systems are universally agreed to be satisfactory, the accountant should be prepared, on behalf of his managing director, to follow up the questions raised by variance analysis when these demand answers. Equally, the production manager should do his utmost to see that the questions are answered before the accountant needs to follow up, even raising questions of his own with the accountant when necessary.

Members of the management team, therefore, obviously need to use a common language and technical terms, accounting or otherwise, must be clearly understood if they are to be included in the reporting vocabulary.

17

The user of financial statements

FORMAT

The design of financial statements should naturally be governed by the requirements of their users; group chairmen, departmental managers, shareholders, banks, tax inspectors and employees. Each user category has its own standpoint and a common set of financial statements will, therefore, be of limited value.

A group chairman is paid to take a broader view and will inevitably require less detailed information regarding the departmental performance of a member company than, say, one of its departmental managers. The chairman may very well be interested in departmental performance in total, leaving the managing director and his managers to concern themselves with, and report on, achievement at this level. In fact the managing director will be less involved with the department's activity in detail, even though he may be more aware of it than the manager.

The chairman and higher management generally, then, should be supplied with summaries incorporating departmental totals, thus obviating the need to read through unnecessary paperwork, the natural corollary being that the statements issued to lower management levels require to be much more detailed. These managers, of course, would normally only receive such detailed statements as relate to their own spheres of influence.

The shareholders' interest centres around the viability of the company in terms of preservation of capital and its earning, and therefore dividend paying, capacity. The annual audited accounts and reports serve as the chief medium by which the shareholder is able to assess both the soundness of his investment and the stewardship of the directors.

Audited accounts are typically designed in such a way that certain key data for the financial year in question are readily ascertainable. These include:

(a) the value of fixed assets of each category and how much has been provided for depreciation;
(b) current assets broken down into stocks, debtors and cash at bank and in hand;
(c) current liabilities, i.e. creditors and accrued charges, provision for taxation, bank overdraft, and any proposed dividend;
(d) share capital plus any capital reserves;
(e) revenue reserves, summarising changes since the previous audited accounts;
(f) sales for the year and profit before tax, and
(g) the charges for taxation and dividends, and the amount of retained profit for the year.

Some changes in presentation in certain cases would appear to follow from proposed legislation giving effect to the EEC's Fourth Directive, one result of which will be to introduce simpler requirements for small companies. Small companies are defined as not having more than one of: (a) sales in excess of £1.4 million; (b) net assets in excess of £700,000, and (c) 50 or more employees.

Profit and loss accounts and directors' reports will not require to be filed and only an abridged version of the balance sheet is to be mandatory.

Valuable though audited accounts are in many respects, they do not greatly help management in the day-to-day running of the business.

In the same way neither would financial statements be of use to banks, tax inspectors or employees and a brief word or two as to the requirements of these user categories will give some indication as to why different layouts are necessary.

Banks are often approached for loans of considerable sums of money and to enable appropriate consideration to be given, will call for up-to-date information as to the financial standing of the applicant. This may well include a balance sheet subsequent to the latest published accounts, together with order book details, cash flow and profit projections, details of insurances and any charges on assets which may have been given as security.

The audited accounts by themselves are inadequate too for tax purposes and detailed tax computations are therefore necessary. In the case of many smaller and medium-sized companies these are prepared by the auditors.

The Employment Protection Act 1975 embodies provisions for the disclosure of financial information by an employer if requested by a trade union. There are certain safeguards, one of which is that the employer is not required to disclose information which would cause substantial injury to his undertaking, and exemption would normally, therefore, be sought in respect of commercially sensitive information. Employee statements are often based on the added value concept whereby all other elements, described as the added value, are added to the cost of materials purchased. It could be stated, for example, that for each £1 spent on purchases there was an added value of X. The use of employee statements will almost certainly become more widespread and the format is destined to be the subject of considerable experimentation and refinement.

Our main concern here, however, is factory management and it may be helpful to illustrate a typical financial statement's distribution system by using the sample management accounts from Appendix 4:

Page no.

1	Balance sheet
2	Fixed assets
3	Cash flow statement
4	Debtor
5	Stocks and work in progress
6	Profit and loss summary
7	Manufacturing overhead
8	Design and development overhead
9	Selling overhead
10	Administration overhead
11	Order book

Distribution:

Page no.

1, 6, 11	Group chairman
1–11	Managing director
1–3, 5–7, 11	Production manager
1–3, 6, 8, 11	Chief engineer

1–6, 9, 11   Sales manager
1–11   Company secretary

As mentioned in the Appendix notes, subsidiary data would also be issued and lower level management would receive copies, as appropriate, e.g. daily non-productive labour times by code number to, say, a machine shop supervisor and his manager.

The management accounts themselves would normally be accompanied by the accountant's report, concentrating on the financial highlights and there may also be managers' reports submitted to the managing director, who in turn would submit a business report to his chairman.

INTERPRETATION OF FINANCIAL STATEMENTS

Managers owe a duty to themselves and to the company to possess a sufficient level of numeracy so as to understand and interpret the financial statements with which they are issued.

All such statements carry a message of some kind and it is of supreme importance that the message be interpreted into action when the occasion demands. For example, a subsidiary report frequently used as a management tool in monitoring labour utilisation, could be called a payroll report, a direct labour summary, or some other name. Depending on the type of business, it is usually possible to reduce to a handful of key factors those parts on which the manager should concentrate his attention: e.g. (a) the number of operators, (b) the hourly output rate, and (c) the number of non-productive hours worked.

The output rate should be as high as possible and the non-productive hours as low as possible, but it is often forgotten that, whilst both these criteria may be met, if the number of operators is seriously below plan then production may be insufficient to recover overheads fully.

As an aid to interpretation, some notes are given on the sample management accounts in Appendix 4, by way of illustration.

Page 1 Balance sheet
Although satisfactory in many respects, with net worth having increased by £35,709 in the half-year, there has been a decrease in liquidity and a careful watch would need to be kept to see that serious overstocking is avoided, and that the timing of invest-

ment in fixed assets is strictly controlled in order to conserve cash and contain current liabilities.

Page 2 Fixed assets
Does not call for special comment, but should be compared with the capital budget and, in the light of the balance sheet comments, capital commitments may need to be reviewed.

Page 3 Cash flow statement
Generally satisfactory, with a cash inflow of £27,350, compared with a budgeted outflow of £60,000.

Page 4 Debtors and current liabilities
Noteworthy that of the sales ledger total of £168,906, 58 per cent is represented by accounts outstanding for one month or more, compared with 42 per cent of the purchase ledger total. This suggests that steps may need to be taken to bring customers' accounts into line with normal credit terms.

Page 5 Stocks and work in progress
Net stocks have increased in the half-year by £64,838, with a significant proportion of this increase occurring in finished equipment and service parts stock, i.e. those categories most readily capable of being turned into sales. A check with sales management will determine whether sales department is aware of the particular models, components etc. involved and whether these can be substantially sold at an early date.

Page 6 Profit and loss summary
Some ground lost, with current month profit of £6,240 falling £3,460 short of budget, although the cumulative profit of £35,709 is still £8,209 better than budget. Chief factors in the current month shortfall were sales, £3,960 below budget and gross margin at 50.3 per cent of sales, instead of 55.0 per cent budgeted. Had the budgeted gross margin percentage been attained, gross margin would have been increased by £3,198.

Cumulatively, although all product groups are showing higher than budgeted gross margin percentages, unbudgeted excess costs of £12,600 account for 3.1 per cent of sales, reducing the total gross margin percentage virtually to that budgeted.

Page 7 Manufacturing overhead
The reason why indirect payroll costs are running over budget requires explaining and expenditure on consumable supplies and loose tools needs to be kept in check.

Page 8 Design and development overhead
Here, indirect payroll costs show a saving. Is this the result of a

budgeting error? If caused by unfilled vacancies, is the department's functioning likely to be impaired?

Page 9   Selling overhead

As with all other departments, occupancy costs are running over budget. This requires explanation. Apart from this and the administration overspend, this overhead statement should not cause undue concern.

Page 10   Administration overhead

Indirect payroll costs should be watched. Of the cumulative excess of £1,566, £962 has arisen in the current month. Cumulative administration expenses include £4,651 in respect of adjustments to prior-year accounts, which would not have been budgeted. Eliminating this item reduces the administration overspend to £1,170.

Page 11   Order book

Cumulatively, orders received at £417,100 are slightly ahead of invoiced sales, £401,900. However, the order book shows a reduction in respect of factored products and service parts and it now becomes obvious that, referring to page 5, service parts are overstocked, with stocks at cost £26,335 representing something like £50,000 sales value. Can the order intake be increased, particularly for these two items?

INTERNAL AND EXTERNAL YARDSTICKS

The key yardstick internally is unquestionably the budget. This remains so, even though it is generally accepted that it may well become out of date in certain aspects, giving rise to the need for modified forecasts for such portions as may be appropriate if it is thought that a full-scale revision is unnecessary.

It is perfectly right that the budget should be given such a high degree of importance. Considerable thought and time will have been invested in its compilation and agreement will have been reached by management, the plan will have been published and then managers will be anxious to compare performance with plan.

A number of subsidiary yardsticks will have been created in the process of constructing departmental and company budgets. These may include:

(a) the qualifying output levels in relation to incentive payment

schemes and average operator efficiency ratings;
(b) departmental standard hourly pay rates and headcount totals;
(c) machine utilisation and allowable downtime;
(d) average throughput times for certain product ranges;
(e) weekly materials issues totals for process operations, and
(f) materials standard prices.

In all these areas and many others, management and supervision should be aware of how well or badly performance compares with plan. It is sometimes instructive to look back to past performance in earlier years and this can, indeed, place the situation in perspective on occasions. Past performance, by definition, would not normally be used as an overriding yardstick in a forward looking organisation.

There is a wealth of information published by government and trade sources from which external yardsticks can be selected. Among the most useful of these data are interim comparisons, which have grown considerably over the years and are now available for most sectors of industry.

Participants in such schemes are generally assured confidentiality and receive detailed reports listing key factors and their statistics. From these management can assess the firms' performance in relation to the trade in general and in what respects improvements need to be made. Detailed tables are available giving financial and operating ratios, production costs and labour productivity. Comparability is obtained usually by providing participants with full sets of instructions and definitions of terms used and supplying a questionnaire for completion.

An interfirm comparison issued to a group of participating foundries showed (a) the number of participants, and (b) the cost of molten metal per 100 kilograms, broken down into wages and 12 other expense headings, quoting three figures for each item: the lowest cost, the highest cost and the average for the group.

This type of information has been known to help managers avoid becoming complacent when internal yardsticks have been met, or bettered, only to find that the firm is below average for the trade sector concerned.

RATIOS AND GRAPHICAL PRESENTATION

The presentation of financial operating data for factory management merits careful attention. Whilst content must take pride of place, the manner in which it is displayed is very nearly as important.

It is instructive to present certain selective statistics in the form of ratios often expressed as percentages. The most commonly used are:

1 Operating profit to capital employed.
2 Operating profit to sales.
3 Net fixed assets to total net assets.
4 Cost of sales to stocks.
5 Liquid assets to current liabilities.
6 Overdue sales ledger accounts to total sales ledger.

Reverting again to Appendix 4, these ratios can be calculated from the sample management accounts illustrated.

1 Operating profit £35,709; annualised $= £71,418 = 20.1$ p.c.
   Capital employed (average) annualised $= \overline{£355,709}$

2 $\dfrac{\text{Operating profit}}{\text{Sales}}$

   This percentage is already shown in the profit and loss summary.

3 $\dfrac{\text{Net fixed assets} = £195,395}{\text{Total net assets} = £355,709} = 54.9$ per cent

   This statistic reveals how much of the firm's capital is invested in fixed assets. Too high a percentage could imply insufficient capital available for current financing.

4 Cost of sales (sales less manufacturing profit) $= £297,535$;
   $\dfrac{\text{annualised} = £595,070}{\text{Stocks} = £297,750} = 2\colon1$

   The company is only turning over its stock twice a year, suggesting overstocking. The notes on the order book earlier in this chapter certainly confirm this in the case of service parts.

   Sometimes stocks are reported in terms of the number of weeks forward sales at cost.

5 $\dfrac{\text{Liquid assets} = £233,416}{\text{Current liabilities} = £216,804} = 1.08\colon1$

   The aim should be to preserve at least a 1:1 ratio to avoid possible delays in meeting commitments, which in turn could lead to production hold-ups. See earlier notes on the balance sheet.

6   Overdue sales ledger accounts
    Total sales ledger
        The relevant percentage has already been quoted in the
    notes on debtors' and current liabilities. An alternative
    statistic is the number of days' sales represented by trade
    debtors' balances outstanding, though it would probably be
    necessary to allow for the influence of VAT.

If statistics, such as those outlined, are plotted monthly, this can be
useful in highlighting trends and focusing attention on problem areas.

Although bar charts and other forms of pictorial presentation, such
as the 'cake' diagram, with which most readers will be familiar, can be
useful as alternative ways of displaying the breakdown of income into
the various cost components plus profit, graphs are probably more
commonly employed in reporting financial data for factory and general management control. The four graphs shown below may be
helpful in demonstrating how such information can be presented and
it will be appreciated that many more could be selected.

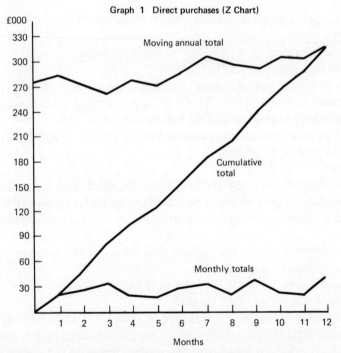

Graph 1   Direct purchases (Z Chart)

Note: The moving annual total is arrived at each month by totalling the
        immediate past 12 months ' figures.

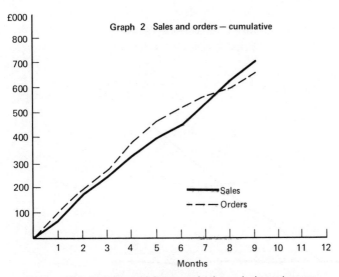

Graph 2   Sales and orders — cumulative

Notes   (1) In businesses with longer production cycles it may be more
              informative to plot 24 months' figures.
          (2) This illustration shows sales starting to outpace order intake.
              Management will need to see that order intake does not lag
              too seriously behind sales to ensure that production gaps are
              not created.

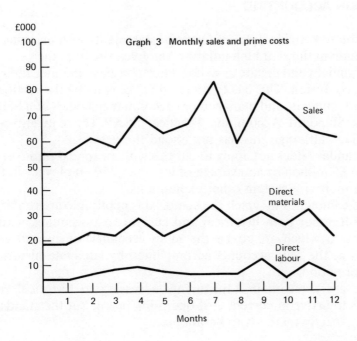

Graph 3   Monthly sales and prime costs

131

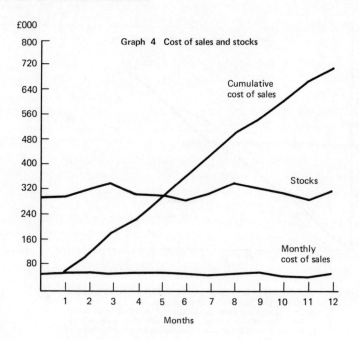

Graph 4 Cost of sales and stocks

INFLATION ACCOUNTING

The desire to provide users of financial statements with meaningful information in times of high inflation has given rise over the years to much argument and debate as to how best to achieve this aim without creating confusion. This has culminated in the issue by the certified, chartered, and cost and management accountancy bodies of a Statement of Standard Accounting Practice, SSAP 16, 'Current cost accounting', although this has not ended the debate.

The standard does not apply to entities with an annual turnover of less than £5 million or an average of less than 250 employees in the UK and there are certain other exemptions.

Those companies to which the standard is applicable are required to publish either both historical and current cost accounts, with a choice as to which are to be the main accounts, or current cost accounts as the only accounts accompanied by adequate historical cost information.

There are three main adjustments necessary to historical cost accounts to arrive at current cost operating profit, but the standard does not specify methods to be adopted.

1 Depreciation adjustment. This is defined as being 'the difference between the value to the business of the part of fixed assets consumed during the accounting period and the amount of depreciation charged on an historical cost basis'.
2 Cost of sales adjustment. 'The difference between the value to the business of stock consumed and the cost of stock charged on an historical cost basis.'
3 Monetary working capital adjustment. Represents 'the amount of additional, or reduced, finance needed for monetary working capital as a result of changes in the input prices of goods and services used and financed by the business'.

The value to the business is, generally, net current replacement cost and monetary working capital is generally taken to be the net total of trade debtors and prepayments, on the one hand, and trade creditors and accruals on the other. To this latter, however, may need to be added fluctuations in bank balances caused by stock and monetary working capital volume changes.

Although paragraph 32 of SSAP 16 explains that it is not a system of accounting for general inflation, paragraph 4 makes clear that, as opposed to historical cost accounts, current cost accounts are designed to reflect changes in input prices of goods and services used and financed by the business affecting the amount of funds required to maintain the operating capability of the net operating assets.

An appendix to the standard gives examples of the presentation of current cost accounts, and the greatest care must be taken when computing appropriate adjustments (see Chapter 15, 'Depreciation').

There is general agreement even among opponents to the standard, on the capital maintenance concept which it embodies, and the continuing debate centres mainly around the need, or otherwise, of a different system of accounting to attain this aim.

An earlier provisional statement, SSAP 7, 'Accounting for changes in the purchasing power of money', dealing with current purchasing power accounting and replacement cost, was withdrawn.

MARGINAL COSTING

Provided that it is not used indiscriminately, for reasons which will become obvious, the presentation of marginal cost data is frequently an extremely worthwhile management decision-making aid.

Marginal cost has been defined as the amount at any given volume of output by which total costs are altered if the output volume is changed by one unit. This has as its basis, the segregation of costs into fixed and variable types, according to whether or not they vary in proportion to the activity level.

Direct labour is usually regarded as a variable cost, although it is recognised nowadays that, primarily as a consequence of employment protection legislation, in the short term at least it is fixed. In the long term, fixed overheads may become variable to some extent. A permanent increase or reduction in the direct labour force could result in more or fewer supervisors being required.

As long as they are recognised, however, these reservations need not invalidate the marginal costing approach as an alternative to total, or absorption, costing in particular circumstances.

One area in which marginal costing techniques have been employed is that of calculating break-even sales levels and prices. A conventional break-even chart is illustrated (p. 135), showing sales of 80,000 units needed to break-even.

Sometimes an estimated break-even figure is included in the profit and loss statement of the management accounts for information, albeit usually somewhat crude and normally assuming an unchanged sales mix.

Referring again to the profit and loss summary in Appendix 4, and taking the manufacturing profit line, it can be seen that the actual half-year total of £104,392 equalled 26 per cent of sales. Dividing total other overhead of £68,683 by 26 per cent or 0.26 produces a break-even sales figure of £264,165. In this instance the exercise is perhaps of only academic interest, as actual turnover was £401,927. Had sales only reached, say £230,000 however, management might well have liked to have had a rough idea of the additional turnover required to break-even. A break-even figure can, of course, be worked out for each column of the statement.

The technique is also used in contribution pricing. Imagine a company with two factories, one making metal cabinets and the other locks and fittings. Each factory has its own works management and production control but both share common sales and administration.

Pricing policy should be to strive for a net contribution sufficient to realise a profit after meeting selling and administration costs, as in the example. It is common practice in this type of situation for each factory to be given a specified percentage mark-up to be applied to factory cost in the calculation of selling prices.

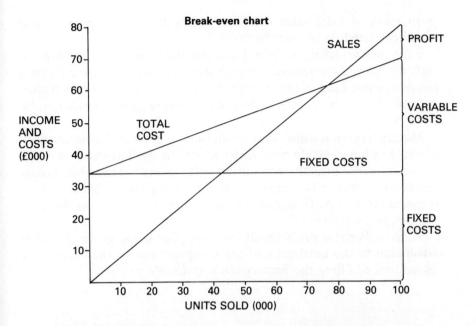

Break-even chart

In cases of surplus capacity, it is conceivable that factory targets may have been met and still more work could be brought in at prices below the norm and as long as marginal costs were covered and a contribution made, additional profit would be generated. The obvious danger of this type of work growing to form an unacceptably high

	Factory A	Factory B	Total
	£000	£000	£000
Sales	100	80	180
Marginal costs	50	45	95
Contribution (gross)	50	35	85
Fixed overhead (factory)	20	25	45
Contribution (net)	30	10	35
Fixed overhead (selling and administration)			20
Profit			15

proportion of total sales must be watched and management will obviously need to be alert to this.

Even when factory targets have not been met, due perhaps to difficult market conditions, contribution pricing can prove to be a tenable proposition as each factory has its own fixed costs which must be paid for, even before selling and administration costs can be covered.

Marginal costing is also a useful method to apply in the comparison of make or buy costs. In most factories only a small proportion of the overhead varies directly with output volume and overhead costs can, therefore, virtually be ignored when assessing the in-house cost of manufacturing a particular component in order to compare this with the bought-out cost.

To do otherwise often results in a supplier being given an unfair advantage to the detriment of the company and works against the broad aim of filling the factory with profitable work.

APPENDICES

APPENDICES

Appendix 1   Sample forms

CONTENTS

Appendices

ESTIMATE SUMMARY											
Enquiry no.	**Customer:**						**Customer ref:**		**Date:**		
Product:				**Part No:**					**Total qty:**		

Delivery period:		Unit costs									
			Prime			Overhead			Total		
	Hrs	Rate	Amount		Rate	Amount		Rate	Amount		
Materials:											
Raw materials			X	X							
Bought out parts			X	X							
Sub-contract			X	X							
Finishing			X	X							
Sub-total			X	X							
Scrap and rework		X%	X	X							
Total			X	X	X%	X	X		X	X	
Labour:											
Department A	X	X	X	X							
Scrap and rework		X%	X	X							
Sub-total			X	X	X%	X	X				
Department B	X	X	X	X							
Scrap and rework		X%	X	X							
Sub-total			X	X	X%	X	X				
Total			X	X		X	X		X	X	
Manufacturing cost			X	X		X	X		X	X	
General administration						X	X	X%	X	X	
Sales and distribution						X	X	X%	X	X	
Company cost			X	X		X	X		X	X	
Warranty								X%	X	X	
Total									X	X	
Profit mark-up								X%	X	X	
Formula selling price									X	X	
Round off to								£	X	X	

Prepared by:		Checked:	
Remarks:			Approved:

ORDER ACKNOWLEDGEMENT

To: Ref: _____

                                                                     Date: _____

We acknowledge with thanks receipt of your purchase order Ref. _____
dated_____ for the supply of the following, subject to the terms and
conditions printed on the reverse of this acknowledgement form

Item No.	Qty.	Product ref.	Description	Price each

Delivery requirements:

For and on behalf of
XYZ Manufacturing Ltd

— — — — — — —

Appendices

	WORKS ORDER		No.		

Customer: Customer Ref. _____   Date _____

Delivery schedule:

Manufacture/ supply of: –

Qty.	Product no.	Description	Sales value	
			Each	Total

Tooling/ test equipt:

Packing:

Delivery address:

Note: A typical distribution scheme might be as follows:-
         Copy 1 (white)   Works Manager
         Copy 2 (yellow)   Production Control
         Copy 3 (pink)   Q.A. Manager
         Copy 4 (green)   Chief Engineer
         Copy 5 (grey)   Materials Control
         Copy 6 (blue)   Accounts

PURCHASE REQUISITION		No:

Dept. ordering _____

Delivery required _____

Supplier (if known):

Date _____

Cost code. _____

Goods:	Price

Departmental Head:	Purchasing sig:	Order No:

GOODS RECEIVED NOTE			No:

Supplier:	Advice note	Order No:	Date:

Item No.	Part No.	Description	Quantities			
			Advised	Rec'd.	Accep'd.	Rejec'd.

G.I.Inspector	Reject note:	Debit note:	Accounts:

Note: Possible G.R.N.distribution:-
Copy 1 (white)   Purchasing
Copy 2 (grey)   Stores via Materials Control (Stock Records)
Copy 3 (blue)   Accounts via Goods Inwards Inspection
Copy 4 (buff)   Goods Inwards

PURCHASE ORDER

XYZ Manufacturing Ltd
North Works
Anytown

Purchase Order No.

Date

To:

Company Reg. No.

England

Vat Reg No.

Your quotation ref.

Please supply the following:-

Item No.	Qty .	Part No.	Description	Price

Subject to the conditions printed overleaf

Pricing basis _____ Signed

Delivery requirements:

_____
Purchasing dept.

Note: Possible distribution:-
          Top copy (white)   Supplier
          Copy 2 (white)   Purchasing
          Copy 3 (buff)   Goods Inwards
          Copy 4 (blue)   Accounts
          Copy 5 (yellow)   Originator

REJECTION NOTE	No.

Supplier:	Date

Supplier's advice note ref.	Order No.

The goods referred to do not conform to
specification for the following reasons:-

Qty.	Part No.	Description

☐ Scrap

☐ Rectify at supplier's cost

☐ Return for rectification

☐ Return for replacement

☐ Despatch to supplier and
charge carriage

☐ Supplier to collect

☐ Credit and re-invoice

As agreed with
supplier

Advice Note No.

_____

Sig. _____ Inspection Dept.	Accounts:
	Debit Note No. _____

Appendices

<table>
<tr><td colspan="8">XYZ Manfg. Ltd.

 STOCK RECORD CARD

</td></tr>
</table>

Order No.	Date	Supplier	Qty.	Price	Standard cost	
					Date	

Unit of measure ☐   Min ☐   Max ☐   Re-order Qty ☐

Receipts				Issues			Bal'ce	Allocations		
Order No.	Goods in. ref	Date	Qty.	Req. No.	Date	Qty		Ref.	Qty.	Free Stock

Drawing/Part No.   Descrip.

XYZ Manfg. Ltd

PARTS LIST

Assembly/component description							Drawing No.		

Works 0/No.		Batch ref			Batch qty				

Item No.	Part No.	Description	Qty per assy	Total batch qty.	Qty issued	Short-age	Comments

Issue No.	Date	Drawing Office			
		Prepared by:		Checked/approved:	

Note: The layout of this form enables it to be used in support of a materials requisition. Parts lists are also frequently prepared in priced form.

MATERIALS REQUISITION

Works order No.		Batch No.	Batch qty.		Drawing No.	

Assembly/component title			Insp. level	Final assy/comp No.		

Op.	Description			Qty	Cost

Shortage Yes ☐ No ☐	Issued by	Received by	Date	Stock records	Costs

Appendices

OPERATION CARD			
Works order No.	Batch No.	Batch qty	Drawing No.
Assembly/component title		Insp. level	Final assy/comp. No.

Op.	Dept.	Description	Tooling/jigs etc.	Mins. ea.

Qty reject	Qty good.	Insp.	Date.	Std. batch time (hrs)			Prodn. Control	Costs
				Set	Run	Total		

OPERATION CARD (REVERSE SIDE)

Date	Clock No.	Times		
		On	Off	Taken

Remarks:

Supervisor's sig. _____

Note: This system allows the times taken to be accumulated on a daily basis , normally by the time clerk and upon the card being stamped 'completed' or even before, supervision can compare actual times with the standard time entered on the face of the card by production control.

PROCESS LAYOUT				
Works order No.	Batch No.	Batch qty.	Drawing No.	
Assembly/component title		Insp. level	Final assy/comp. no.	

Op	Dept.	Description	Tooling/jigs etc.	Mins ea.	Qty. to next op.	Date

Scheduled completion date	Total assy/ Comp. std. Batch time (hrs)	Set Run Total	Prodn. control	Costs		

Note: The process layout, often known as the route card, is frequently issued with the material requisition(s) and operation cards as a multi-3 part set, with a duplicate copy of the final operation card being produced and endorsed 'route to stores' thus serving as the stores receipt for booking in the goods produced.

Appendices

| | SUMMARY TIME SHEET | | | | | W/e _____ |

Dept. _____   Clock No. _____   Name _____

Works order No.	Non-prod. code	Sun	Mon	Tue	Wed	Thu	Fri	Sat	Total hrs	Costs
Initialled ——▶										
Analysis:										

Note: The use of such a form, one for each operator, facilitates agreement of the weekly payroll analysis and can be entered daily by the time clerk from the operation cards and checked with the clock cards. This then enables the operation cards to continue in use until completed, rather than closed weekly, and means that more than one operator can be booked to the same operation card. Can often be used with advantage in short batch production situations.

TEMPORARY OPERATION NOTE NO.	
Works order No:	Drawing No:
Batch No.	Assy/component title:
Batch qty:	

Tick if new operation ☐

Original operation No. (s)
If substituted operation _____

Reason for temporary operation:

Production eng. authorisation:	Allowed time:	Date:

RETURNS INWARDS NOTE NO:			
Customer:	Advice note:	Invoice No:	Date:

Item No.	Part No.	Description	Quantities			
			Advised	Rec'd	Accep'd	Rejec'd

Reason for return:		Sales dept.
Disposal instructions:	G.I. Insp.	Credit note:

Appendices

	SCRAP NOTE		No.

Dept.	Works order	Batch ref.	Date

The following goods have been scrapped
in production:

Part No. _____ Qty. _____

Description:

_____

Last completed operation No: _____

Reason:

_____

Disposal instructions:

Inspection auth.		Costs:	
		Unit	Total
	Matl.		
	Lab.		

152

JOURNAL VOUCHER		No. _____		
		Month_____19___		
Description	Code	Debit		Credit	

Details :

Prepared by: _____

Approved: _____

Appendices

CAPITAL AUTHORISATION		No:

Date of proposed expenditure:	Budget ref:

Details of proposed acquisition:	Asset category	Costs	
		Internal	External

If ancillary to existing plant quote
original asset ref.

Required for _____ Dept.	Supplier:
Location:	

Investment justification (attach supporting data, eg D.C.F.
calculations)

Estimated useful life of asset:

Details of existing assets rendered redundant and
disposal plans:

Warranty/maintenance contract details:

External authority approval required?   Yes ☐   No ☐

Originator:	Dept. head:	Accountant:	Managing Dir:
Date:	Date:	Date:	Date:

Appendix 2   Sample reports

CONTENTS

CAPITAL EXPENDITURE SUMMARY – 1st Qtr.

Ref:	Dept	Budget Description	Cost £	Cap. Auth.	Actual cost Month 1 £	Actual cost Month 2 £	Actual cost Month 3 £	(Over) under Spend £
1	QA	Viewmaster projector	25,000	00182	25,600			(600)
2	Factory 2	Multi-station drill	11,000	00282		9,900		1,100
3	Stores	ABC conveyor system	4,000	00382				4,000
4	Stores	Electron scales	2,000	00482		2,200		(200)
5	Factory 2	Env. extractor plant	17,000	00582				17,000
6	Goods in	S/H fork lift truck	6,000					6,000
7	Sales	Office furniture	1,800				1.750	50
8	Factory 1	Workhorse centre lathe	40,000	00682				40,000
			106,800		25,600	12,100	1,750	67,350

Note: This summary could be issued in partially complete form for Months 1 and 2. All 8 items would be included but the Month 1 issue would omit figures from the last three columns and Month 2 would omit the last two.
The summary may also contain notes amplifying the reasons for budget variances where necessary.

STORES PURCHASES SUMMARY — Month Jul.

Code		Budget	Standard	Price variance	Total
		£	£	£	£
1211-1	Raw materials	5,000	5,315	(495)	4,820
1211-2	Bought out parts	8,000	10,195	1,114	11,309
1211-3	Sub-contract	5,500	7,200	(960)	6,240
1211-4	Finishing	1,500	1,312	256	1,568
		20,000	24,022	(85)	23,937
	Factored products:				
1235-1	Electronic	2,000			3,164
1235-2	Mechanical	4,000			3,608
		6,000			6,772
	Total direct	26,000			30,709
1251	Loose tools and gauges	500			376
1252	Packing materials	300			147
1253	Consumable supplies	1,000			891
	Total indirect	1,800			1,414
	Total stores purchases	27,800			32,123

Note: "Budget" and "Actual" are top-level headings; "Standard", "Price variance", and "Total" fall under "Actual".

Appendices

		Gl. Ref.	Date	Inv. Ref.	Supplier	Amount
						£
2591-1	Raw materials	7095	1.7.81	H234	Blenkinson	33.55
		7128	4.7.81	9238C	Wilkins	(100.98)
		7098	1.7.81	00427	SPO	(22.60)
		7106	2.7.81	6102	Biggs	(43.68)
		7142	4.7.81	11923	Hill & Dale	90.20
		7233	8.7.81	H242	Blenkinson	(116.50)
		7409	15.7.81	H261	Blenkinson	(254.34)
		7293	10.7.81	BT711	BT Company	(147.00)
		7495	17.7.81	00500	SPO	19.40
		7516	20.7.81	2468	Kingmetal	(71.77)
		7556	23.7.81	9346C	Wilkins	24.60
		7609	27.7.81	506	Jakes Steel	85.00
		7650	30.7.81	9398C	Wilkins	7.80

Net (favourable) / unfavourable variance   496.32

Note. The analysis illustrated identifies variances by invoice, enabling ready reference to be made so that significant variances can be more easily investigated and would include similar listings for other purchase groupings. There are, of course, various alternative ways of presenting the information, such as by part number, possibly even including the standard cost of each item for comparison.

STOCK AUDIT — Month Aug.

Weeks 1 to 4

Part No.	Quantities		Discrepancy		Check Sig.	Date	Total Cost (+)—
	Stock record	Actual	Qty.	Unit cost			
02917	12	11	1	1.50	CD	3.8.81	1.50
02907	110	112	(2)	14.00	CD	5.8.81	(28.00)
04133	16	11	5	0.99	CD	5.8.81	4.95
05992	601	598	3	10.47	EF	6.8.81	31.41
06790	12	14	(2)	4.20	EF	7.8.81	(8.40)
46602	20	21	(1)	1.10	LM	8.8.81	(1.10)
46939	4	2	2	16.50	CD	10.8.81	33.00
51803	419	393	26	0.20	CD	12.8.81	5.20
52217	90	95	(5)	9.00	EF	14.8.81	(45.00)
56209	2	1	1	8.25	EF	14.8.81	8.25
99201	64	62	2	20.16	LM	18.8.81	40.32
32001	20	21	(1)	4.00	LM	21.8.81	(4.00)
32324	6	7	(1)	2.20	LM	21.8.81	(2.20)
33090	24	23	1	0.90	CD	21.8.81	0.90
26166	16	10	6	2.50	CD	25.8.81	15.00
28090	44	45	(1)	0.80	CD	27.8.81	(0.80)
66066	62	60	2	10.00	EF	29.8.81	20.00
69117	590	580	10	1.44	EF	29.8.81	14.40
							£85.43

Accounts ref.   JV.19

From: materials control

No. of items checked

In the month   460

To   (1) accounts
      (2) production mangr.

Sig. AB

Note: This is an example of an exception report, there being little point in listing all those parts with no discrepancies. If desired, the report could be issued weekly for information but only totalled at month-end, at which point it could be actioned as a stock adjustment journal voucher.

Appendices

GROSS MARGIN ANALYSIS — Month Aug.

Weeks 1 to 3

	Sales	Materials	% Sales	Labour	% Sales	Gross margin	% Sales
Current week	£	£		£		£	
Cabinets	10,112	4,306	42.6	911	9.0	4,895	48.4
Machines	27,950	9,160	32.8	3,188	11.4	15,602	55.8
Total	38,062	13,466	35.4	4,099	10.8	20,497	53.8
Month to date							
Cabinets	27,805	11,066	39.8	2,341	8.4	14,398	51.8
Machines	98,400	33,890	34.4	10,908	11.1	53,602	54.5
Total	126,205	44,956	35.6	13,249	10.5	68,000	53.9
Budget for month							
Cabinets	60,000	24,000	40.0	4,800	8.0	31,200	52.0
Machines	140,000	49,000	35.0	16,800	12.0	74,200	53.0
Total	200,000	73,000	36.5	21,600	10.8	105,400	52.7

Note: In cases where a standard costing system is in operation, it may suffice for the gross margin analysis to exclude materials and labour variances, although these would, of course, need to be brought into the monthly management accounts.

PAYROLL REPORT — FACTORY 1 — W/E 22.8 19XX

	Hours	Cost			
		Payroll	Transfers	Total	Budget
		£	£	£	£
Productive labour					
Product X	760	1,900		1,900	2,000
Product Y	554	1,385		1,385	1,500
Transfers	168		(420)	(420)	(450)
	1,482	3,285	(420)	2,865	3,050
Variances: Rate		(30)		(30)	
Efficiency		148		148	
		118		118	
Overtime premium					
Supervision	54	149	135	284	275
Rectification	19		48	48	50
Waiting time	12		30	30	20
Training	42		105	105	100
Inspection	8	220	20	240	200
Stores and goods in	11	317	27	344	320
Production control		198		198	225
Other non-productive	22		55	55	40
Employers' NI		587		587	550
	168	1,471	420	1,891	1,780
Total gross pay + NI				4,874	4,830
Net pay				2,880.00	
PAYE				874.50	
NI total				857.20	
Other deductions				260.50	
				4,872.20	

Appendices

NON-PRODUCTIVE LABOUR — W/E 22.8. 19XX

Dept.   Fabrication

Code	Description	Sun	Mon	Tue	Wed	Thu	Fri	Sat	Total
									hrs
NP.1	Supervision		1	2½	2¾				
NP. 2	Rectification		4	2	1½				
NP. 3	Waiting time		½		1				
NP. 4	Training			3	2				
NP. 5	Inspection		1		2½				
NP. 6	Stores and goods inwards			1½					
NP. 7	Meetings			1	1½				
NP. 8	Temporary Ops.		1½		2				
NP. 9	Experimental		2½						
NP. 10	Other non-productive			½					
	Total hours		10½	10½	13¼				
	Supervisor's sig.		A.B.	A.B.	A.B.				

LABOUR EFFICIENCY — W/E 15.8. 19XX

Hours	Dept. 1	Dept. 2	Dept. 3	Dept. 4	Total
Product X					
Standard	158.50		69.50	50.25	278.25
Actual	186.50		90.00	44.25	320.75
Variance	28.00	–	20.50	(6.00)	42.50
Product Y					
Standard	245.00	350.00	22.00	46.00	663.00
Actual	269.50	364.00	15.50	35.50	684.50
Variance	24.50	14.00	(6.50)	(10.50)	21.50
Total					
Standard	403.50	350.00	91.50	96.25	941.25
Actual	456.00	364.00	105.50	79.75	1005.25
Variance	52.50	14.00	14.00	(16.50)	64.00
Cost	£	£	£	£	£
Standard rate	2.50	2.30	2.95	3.40	–
Product X	70.00	–	60.48	(20.40)	110.08
Product Y	61.25	32.20	(19.18)	(35.70)	38.57
Total	131.25	32.20	41.30	(56.10)	148.65
% of standard					
Product X	17.7	–	29.5	(11.9)	15.3
Product Y	10.0	4.0	(29.5)	(22.8)	3.2
Total	13.0	4.0	15.3	(17.1)	6.8

PROJECT COST REPORT – Month Sept.

Works Order No.	Sales value £	Works cost £	Completion month	Materials £	Overhead 10% £	Labour £	Overhead 3,00% £	Works cost £	Sales to date £	Status
WO.1001	80,000	72,500	Aug 81	55,319	5,532	2,395	7,185	70,431	67,500	Overdue materials shortages and cost overrun
WO.1004	45,000	35,800	Nov 81	13,240	1,324	3,482	10,446	28,492	41,000	OK Cost savings
WO.1007	28,500	21,400	Feb 82	6,320	632			6,952	–	Materials procure-ment only
WO.1009	7,800	5,750	Oct 81	2,013	201	744	2,232	5,190	–	Virtually complete
WO.1010	12,000	8,850	Dec 81	2,060	206	930	2,790	5,986	–	Delay – supplier dispute
WO.1011	62,800	47,200	Jan 82	21,320	2,132	2,372	7,116	32,940	34,966	On target
WO.1012	4,300	3,200	Oct 81	1,208	121	160	480	1,969	–	Cancelled – liability claim
	240,400	194,700		101,480	10,148	10,083	30,249	151,960	143,466	

MACHINE UTILISATION – W/E 22.8. 19XX

	Machine No.							Total
	1	2	3	4	5	6	7	
Tape proving		14	5½					19½
Waiting for work				2½			38	40½
Tooling		5				4		9
Tool setting		1						1
Machine breakdown			1½					1½
Cleaning			½	½				1
No operator			2					2
No power				3				3
Awaiting inspection			2					2
Total non-running								
Time (hours)	–	20	11½	6	–	4	38	79½

CASH FORECAST (£000) -- Month Sept. 19XX

Forecast for the month			Actual				
		Week 1	Week 2	Week 3	Week 4	Week 5	Total
27.2	Balance b/fwd	27.2	31.6	29.4			
	Receipts:						
83.5	Sales ledger	28.9	16.9	12.3			
15.0	Progress claims	-	-	6.0			
7.0	Others	1.5	2.3	1.0			
105.5		30.4	19.2	19.3			
	Payments:						
32.4	Purchase ledger	17.3	11.0	3.1			
42.1	Payroll (incl.PAYE)	8.0	6.6	10.5			
8.0	Inter-company	-	-	-			
14.3	Others	0.7	3.8	5.0			
96.8		26.0	21.4	18.6			
35.9	Balance c/fwd	31.6	29.4	30.1			

PRODUCTION INSPECTION REPORT — W/E 22.8 19XX

Works Order No.	Batch Ref.	Total Qty.	Rejected Qty. Total	Scrap	Other	Comments
901	901/1	40	–			OK
903	903/1	28	2	2		Undersize
901	901/2	16	4		4	Burred — reworked
905	905/1	22	1		1	Drill holes oversize / Concession applied for
901	901/3	8	–			OK
906	906/2	66	13	13		Flawed material / Supplier notifed
904	904/1	26	9		9	Faulty wiring — Rework
903	903/3	14	4		4	Plating thickness outside / tolerance — Return to platers
904	904/2	71	3	3		Cracked in machining
		291	36	18	18	

Rejected 12.4%. Scrapped 6.2%

Sig.   BC
_____
Chief Inspector

Appendices

Department A   . No. of employees:

Male   <u>12</u>

Female   <u>16</u>

Hours	<u>Male</u>	<u>Female</u>	<u>Total</u>
Normal basic	468	560	1028
Overtime	<u>54</u>	<u>16</u>	<u>70</u>
Total	<u>522</u>	<u>576</u>	<u>1098</u>
Worked	<u>493</u>	<u>484</u>	<u>977</u>
Lost time:			
Holidays	16	70	86
Sickness	12	16	28
Authorised absence		4	4
Total	<u>28</u>	<u>90</u>	<u>118</u>
Overtime % of normal basic	11.5	2.9	6.8
Lost time % of normal basic	6.2	16.4	11.8

Appendix 3 Sample accounts code

NOTE

Although reasonably comprehensive, this specimen layout can be further extended within the 4-digit capacity provided.

The first digit denotes the particular accounts segment, with the first 500 numbers being pre-fixed 1, i.e. balance sheet accounts allocated within the 1000 to 1499 range.

Alternatively, it will be obvious that the layout provides for the exclusion of any accounts considered unnecessary in particular circumstances.

INDEX

1000 Balance sheet
2000 Sales and prime costs
3000 Manufacturing overhead
4000 Design and development overhead
5000 Selling overhead
6000 Administration overhead
7000 Miscellaneous income and expense

FIXED ASSETS

000   Total fixed assets at cost
    005 Freehold premises and fixtures
    010 Leasehold premises and fixtures

105 Plant, machinery and equipment
020 Tooling
025 Canteen equipment
030 Furniture, fittings and office equipment
035 Commercial vehicles
040 Motor cars
050 Total depreciation provisions
055 Depreciation provision — freehold premises and fixtures
060 Depreciation provision — leasehold premises and fixtures
065 Depreciation provision — plant, machinery and equipment
070 Depreciation provision — tooling
075 Depreciation provision — canteen equipment
080 Depreciation provision — furniture, fittings and office equipment
085 Depreciation provision — commercial vehicles
090 Depreciation provision — motor cars

LIQUID ASSETS

100 Total liquid assets
110 Cash at bank and in hand — total
111 Bank control
115 Petty cash
116 Cash floats
120 Debtors — total
121 Sales ledger
122 Bad debts provision
125 Progress claim debtors
129 Sundry debtors

CURRENT LIABILITIES

150 Total current liabilities
160 Purchase ledger
170 Accrued charges
180 Sundry creditors
190 Provision for taxation

STOCKS AND WORK IN PROGRESS

200		Total stocks and work in progress
	210	Production stores — total
	211	Direct materials and sub-contract charges
	212	Direct labour
	213	Production overhead
	220	Work in progress — total
	221	Direct materials and sub-contract charges
	222	Direct labour
	223	Production overhead
	230	Finished equipment stocks — total
	231	Product A
	232	Product B
	235	Factored products
	240	Service parts stocks — total
	241	Product A
	242	Product B
	250	Sundry stocks — total
	251	Loose tools and gauges
	252	Packing materials
	253	Consumable supplies
	259	Stock provisions

PREPAYMENTS

280	Total prepayments

DEFERRED INCOME AND LIABILITIES

300		Total deferred income and liabilities
	310	Government capital grants
	320	Progress claims suspense
	330	Deferred warranty income
	340	Deferred taxation
	350	Long-term loans

INTER-COMPANY ACCOUNTS

400		Total inter-company accounts
	410	Holding company
	420	Other group accounts — total
	421	Company 1
	422	Company 2
	423	Company 3

CAPITAL AND RESERVES

450		Total capital and reserves
	460	Share capital
	470	Capital reserves
	480	Profit and loss account

SALES AND PRIME COSTS

500		Total sales
	505	Sales — Product A
	510	Sales — Product B
	515	Sales — factored products
	520	Sales — service parts, Product A
	525	Sales — service parts, Product B
550		Total prime costs
	555	Materials cost of sales — Product A
	556	Materials cost of sales — Product B
	558	Materials cost of sales — factored products
	560	Materials cost of sales — service parts, Product A
	561	Materials cost of sales — service parts, Product B
	570	Labour cost of sales — Product A
	571	Labour cost of sales — Product B
	590	Excess costs — total
	591	Materials price variance
	592	Materials scrap variance
	594	Labour rate variance
	595	Labour efficiency variance
	596	Labour scrap and rework variance
	599	Stock adjustments

INDIRECT EXPENSES – PAYROLL

600		Total indirect expenses — payroll
	601	Directors' salaries
	602	Management salaries
	603	Non-productive labour
	604	Holiday pay
	605	Severance payments
	606	Other indirect wages and salaries
	611	Company national insurance
	612	Company pension
700		Total direct expenses — other
	705	Advertising and exhibitions — total
	706	Advertising
	707	Exhibitions
	710	Travel and entertaining — total
	711	Home entertaining
	712	Export entertaining
	713	Foreign travel
	714	Company car expenses
	715	Employee car allowances
	716	Other travel
	725	Carriage and packing — total
	726	Company transport
	727	Carriage — other
	728	Packing materials
	730	Energy costs — total
	731	Electricity
	732	Gas
	733	Fuel oil
	735	Repairs and renewals — total
	736	Repairs and renewals — plant
	737	Repairs and renewals — furniture, fittings and office equipment
	740	Consumable supplies and loose tools
	744	Postage
	745	Printing and stationery
	746	Telephone
	750	Bank charges and interest
	751	Hire purchase interest
	755	Insurances

756 Bad debts
757 Discounts, net
770 General expenses — total
771 Canteen
772 Welfare and training
773 Cleaning
774 Subscription and donations
775 Staff recruitment
779 Miscellaneous

ESTABLISHMENT EXPENSES

800   Total establishment expenses
820 Occupancy costs — total
811 Rent
812 Rates (including water)
813 Depreciation — freehold premises and fixtures
814 Depreciation — leasehold premises and fixtures
815 Repairs — buildings and fixtures
820 Equipment rental
830 Depreciation (excluding buildings) — total
831 Depreciation —plant, machinery and equipment
832 Depreciation — tooling
833 Depreciation — canteen equipment
834 Depreciation — furniture, fittings and office equipment
835 Depreciation — commercial vehicles
836 Depreciation — motor cars

MISCELLANEOUS EXPENSE/INCOME

900   Total miscellaneous expense/income
910 Industrial training levy
919 Other miscellaneous expense
930 Profit/loss on disposal of fixed assets
935 Prior year adjustments
940 Government grants — capital
941 Government grants — other
949 Other miscellaneous income

Appendix 4 Sample management accounts

NOTES

The presentation of the sample management accounts which follow is, of course, intended merely to give an indication of what might be an acceptable layout, according to circumstances.

There are eleven pages in this particular set, with pages 1 to 5 relating to the balance sheet, and 6 to 10 to the profit and loss account, with page 11 summarising the business position in terms of customers' orders. The information presented can be expanded or condensed, as desired, provided that the message being conveyed is not lost sight of in a welter of data or, conversely, that the information is not too scanty for a meaningful message to be received.

Too much attention should not be paid to the figures in themselves, the main object obviously being to illustrate their relationships in constructing the balance sheet and profit and loss account.

In most cases subsidiary data would also be issued, usually in advance of the management accounts, such as payroll reports, weekly, or even daily, sales reports, and detailed sales and costs analyses.

WXYZ Manufacturing Ltd

BALANCE SHEET: PERIOD 6 , 1981/82

Employment of capital:

30.6.1981			
156,380	Net fixed assets		195,395
	Liquid assets:		
17,060	Cash	46,410	
201,207	Debtors	187,006	
218,267		233,416	
198,564	Current liabilities	216,804	
19,703	Net liquid assets	16,612	
232,912	Stock and work in progress	297,750	
252,615	Net current assets		314,362
5,710	Prepayments		3,402
414,705			513,159
34,690	Deferred income and liabilities		42,250
380,015			470,909
60,015	Due to group companies		115,200
£320,000	Total net assets		£355,709

Capital employed:

Share capital:

225,000	225,000 ordinary shares of £1 each (authorised £250,000)		225,000
	Profit and loss account:		
95,000	Balance 30.6.1981	95,000	
–	Add 1.7.1981/30.12.1981	35,709	
			130,709
£320,000	Total net worth		£355,709

177

Appendices

FIXED ASSETS: PERIOD 6, 1981/82

Cost:

	30.6.1981	Additions	Disposals	31.12.1981
	£	£	£	£
Leasehold premises and fixtures	41,000			41,000
Plant, machinery, and equipment	162,196	37,525		199,721
Tooling	8,700	1,400		10,100
Canteen equipment	2,420			2,420
Furniture, fittings and office equipment	20,250	2,060		22,310
Commercial vehicles	7,780			7,780
Motor cars	18,250	8,400	(6,250)	20,400
Total cost	260,596	49,385	(6,250)	303,731

Accumulated depreciation :

	30.6.1981	Additions	Disposals	31.12.1981
	£	£	£	£
Leasehold premises and fixtures	20,500	980		21,480
Plant, machinery and equipment	49,533	3,820		53,353
Tooling	6,200	1,100		7,300
Canteen equipment	800	120		920
Furniture, fittings and office equipment	12,340	900		13,240
Commerical vehicles	2,593	1,300		3,893
Motor cars	12,250	2,150	(6,250)	8,150
Total depreciation	104,216	10,370	(6,250)	108,336
Net totals	156,380	39,015	−	195,395

CASH FLOW STATEMENT: PERIOD 6, 1981/82

| Current month | | | Cumulative | |
Budget	Actual		Budget	Actual
(10,000)	(8,400)	Capital expenditure	(60,000)	(49,385)
1,750	1,750	Depreciation	10,500	10,370
		Movements:		
(30,000)	22,900	Debtors	(20,000)	14,201
4,000	12,693	Current liabilites	–	18,240
(15,000)	(32,729)	Stocks and work in progress	(55,000)	(64,838)
–	112	Prepayments	–	2,308
–	1,420	Deferred income and liabilities	5,000	7,560
31,750	14,194	Group companies	32,000	55,185
7,500	6,240	Pre-tax profit	27,500	35,709
£(10,000)	£18,180	Net inflow/(outflow)	£(60,000)	£29,350

(30,000)	27,230	Opening cash	20,000	17,060
(40,000)	46,410	Closing cash	(40,000)	46,410
£(10,000)	£19,180	Cash balance increase / (decrease)	£(60,000)	£29,350

DEBTORS: PERIOD 6, 1981/82

Sales ledger :

Current month	71,248
1 Month	46,091
2 Month	29,820
3 Months and over	21,747
	168,906 (A)
Less bad debts provision	5,400
	163,506
Progress claim debtors	20,000
Sundry debtors	3,500
	£187,006

(A) Average month's credit   2.6

    Last Month   3.0

Note: Settled in following 7 days £38,350

CURRENT LIABILITIES: PERIOD 6, 1980/81

Purchase ledger :

Current month	70,581
1 Month	49,762
2 Months and over	2,040
	122,383 (B)
Accrued charges	29,309
Sundry creditors	17,612
Provision for taxation	47,500
	£216,804

(B) Average month's credit   1.8

    Last month   1.4

Note: Settled in following 7 days £39,116

STOCKS AND WORK IN PROGRESS: PERIOD 6, 1981/2

	Balance	Input	Output	Stock Adjustments	Balance 31.12.81	Budget
	£	£	£	£	£	£
Product stores	110,449	140,769	(104,530)		146,688	
Work in progress	73,601	125,971	(156,441)	12,400	65,531	
Finished equipment	97,216	329,125	(276,295)	(19,700)	130,346	
Service parts stock	12,796	34,779	(21,240)		26,335	
Sundry stocks	6,700				6,700	
	300,762	630,644	(558,506)	(7,300)	375,600	362,850
Stock provisions	67,850			10,000	77,850	77,850
	232,912	630,644	(558,506)	17,300	297,750	285,000

No. of weeks stock (forward sales at cost)   16.5   15.0

PROFIT AND LOSS SUMMARY: PERIOD 6, 1981/82

		Current month					Cumulative		
Budget	% SALES	Actual	% SALES		Budget	% SALES	Actual	% SALES
£		£		Sales:	£		£	
39,000	54.2	32,101	47.2	Product A	210,000	55.1	220,537	54.9
20,800	28.9	23,390	34.4	Product B	98,000	23.4	93,408	23.2
7,200	10.0	7,119	10.5	Factored products	45,800	12.0	52,292	13.0
5,000	6.9	5,430	7.9	Service parts	36,000	9.5	35,690	8.9
72,000	100.0	68,040	100.0	Total sales	389,800	100.0	401,927	100.0
				Gross margin:				
27,200	69.7	20,551	64.0	Product A	135,000	64.3	143,743	65.2
9,500	45.7	11,169	47.8	Product B	45,000	50.6	55,001	58.9
1,400	19.4	1,410	19.8	Factored products	7,100	15.5	9,028	17.3
1,500	30.0	3,090	48.1	Service parts	13,900	38.6	17,444	48.9
39,600	55.0	36,220	53.2		201,000	52.8	225,216	56.0
-	-	1,976	2.9	Excess costs	-	-	12,600	3.1
39,600	55.0	34,244	50.3	Total gross margin	201,000	52.8	212,616	52.9
18,400	25.6	17,243	25.3	Manufacturing overhead	101,700	26.7	108,224	26.9
21,200	29.4	17,001	25.0	Manufacturing profit	99,300	26.1	104,392	26.0
				Other overheads:				
3,700	5.1	3,420	5.0	Design and development	30,300	8.0	27,150	6.8
7,800	10.8	7,341	10.8	Selling	41,500	10.9	41,533	10.3
11,500	15.9	10,761	15.8	Total other overhead	71,800	18.9	68,683	17.1
9,700	13.5	6,240	9.2	Pre-tax net profit	27,500	7.2	35,709	8.9

MANUFACTURING OVERHEAD: PERIOD 6, 1981/82

Current month			Cumulative	
Budget	Actual		Budget	Actual
£	£	Indirect manufacturing expenses:	£	£
7,200	7,783	Indirect payroll	44,000	46,091
200	152	Travel and entertaining	1,400	1,600
800	690	Energy costs	5,000	4,634
750	793	Repairs and renewals	4,500	4,939
500	610	Consumable supplies and loose tools	3,000	4,595
200	190	Telephone	1,500	1,393
1,150	924	General expenses	6,900	6,960
10,800	11,142		66,300	70,212

		Establishment expenses:		
3,250	3,200	Occupancy costs	21,150	21,370
50	–	Equipment rental	350	400
750	750	Depreciation – plant and tools	4,500	4,920
4,050	3,950		26,000	26,690

		Miscellaneous expenses (income):		
250	250	Industrial training levy	1,400	1,400
(50)	(100)	Government grants	(400)	(800)
200	150		1,000	600
2,750	3,211	Administration allocation	16,600	19,647
17,800	18,543	Total manufacturing overhead	109,900	117,149
600	(1,210)	Work in progress overhead change	(8,200)	(8,925)
18,400	17,243		101,700	108,224

183

DESIGN AND DEVELOPMENT OVERHEAD: PERIOD 6, 1981/82

Current month			Cumulative	
Budget	Actual		Budget	Actual
£	£	Indirect design and development expenses:	£	£
2,100	1,806	Indirect payroll	18,600	14,862
40	18	Travel and entertaining	1,000	907
60	54	Energy costs	500	394
40	–	Repairs and renewals	300	117
70	25	Consumable supplies and loose tools	600	283
40	32	Telephone	250	197
50	73	General expenses	500	519
2,400	2,008		21,750	17,279
		Establishment expenses:		
300	310	Occupancy	2,000	2,420
50	–	Equipment rental	300	180
350	310		2,300	2,600
950	1,102	Administration allocation	6,250	7,271
3,700	3,520	Total design and dev. overhead	30,300	27,150

SELLING OVERHEAD: PERIOD 6, 1981/82

| Current month | | | Cumulative | |
Budget	Actual		Budget	Actual
£	£	Indirect selling expenses:	£	£
3,000	2,630	Indirect payroll	17,750	16,628
1,200	1,036	Advertising and exhibitions	2,400	2,619
370	396	Travel and entertaining	2,000	1,896
250	221	Carriage and packing	1,600	1,530
80	79	Energy costs	600	482
40	28	Repairs and renewals	250	301
100	114	Telephone	750	636
10	36	General expenses	100	298
5,050	4,540		25,450	24,390
		Establishment expenses:		
350	410	Occupancy	2,400	3,040
50	100	Equipment rental	400	600
400	400	Depreciation commercial vehicles and cars	2,800	2,800
800	910		5,600	6,440
		Miscellaneous expense/ (income) :		
–	–	Profit on sale of motor car	–	(1,500)
1,950	1,891	Administration allocation	10,450	12,203
7,800	7,341	Total selling overhead	41,500	41,533

Appendices

ADMINISTRATION OVERHEAD: PERIOD 6, 1981/82

Current Month			Cumulative	
Budget	Actual		Budget	Actual
£	£	Indirect administration expenses:	£	£
2,700	3,662	Indirect payroll	14,300	15,866
70	76	Travel and entertaining	500	414
120	99	Energy costs	800	634
30	18	Repairs and renewals	250	381
180	206	Postage	1,250	1,327
350	294	Printing and stationery	2,500	2,693
120	101	Telephone	1,000	704
400	–	Bank charges and interest	1,500	–
550	550	Insurances	3,500	3,861
350	350	Bad debts	2,500	3,196
–	(18)	Discounts, net	–	(108)
30	42	General expenses	180	212
4,900	5,380		28,280	29,180
		Establishment expenses:		
480	554	Occupancy	3,250	3,560
10	10	Equipment rental	100	60
260	260	Depreciation	1,670	1,670
750	824		5,020	5,290
		Miscellaneous expense/(Income):		
–	–	Prior year adjustments	–	4,651
5,650	6,204		33,300	39,121
		Allocation:		
2,750	3,211	Manufacturing	16,600	19,647
950	1,102	Design and development	6,250	7,271
1,950	1,891	Selling	10,450	12,203
5,650	6,204		33,300	39,121

ORDER BOOK: PERIOD 6, 1981/82   (£000)

Current month	Product A	Product B	Factored products	Service parts	Total
Opening order book	239.8	98.7	25.5	19.6	383.6
Orders received	45.5	35.4	12.6	10.6	104.1
	285.3	134.1	38.1	30.2	487.7
Sales invoiced	32.1	23.4	7.1	5.4	68.0
Closing order book	253.2	110.7	31.0	24.8	419.7

Cumulative	Product A	Product B	Factored products	Service parts	Total
Opening order book	229.9	101.6	46.0	27.0	404.5
Orders received	243.8	102.5	37.3	33.5	417.1
	473.7	204.1	83.3	60.5	821.6
Sales invoiced	220.5	93.4	52.3	35.7	401.9
Closing order book	253.2	110.7	31.0	24.8	419.7

Index

189